Readings in Environmental Psychology

Perceiving Environmental Risks

Readings in Environmental Psychology
Series Editor
David Canter
University of Liverpool, U.K.

Readings in Environmental Psychology

Series Editor

David Canter

Perceiving Environmental Risks

Edited by

Tim O'Riordan

School of Environmental Sciences, University of East Anglia, U.K.

ACADEMIC PRESS

Harcourt Brace & Company, Publishers

London   San Diego   New York   Boston   Sydney   Tokyo   Toronto

ACADEMIC PRESS LIMITED
24–28 Oval Road
London NW1 7DX

United States Edition published by
ACADEMIC PRESS INC.
San Diego, CA 92101

This book is printed on acid-free paper

A catalogue record for this book is available from the British Library

ISBN 0-12-527840-3

Typeset by Galliard (Printers) Ltd, Great Yarmouth, Norfolk
Printed and bound in Great Britain by Hartnolls Ltd, Bodmin, Cornwall

CONTENTS

PREFACE

Ask most non-specialists about 'the environment' and what it means and the chances are that they will answer in terms of damage and degradation, destruction and pollution, probably at a global scale. What they are responding to is their perception of the dangers and risks associated with the large-scale environment. Given the scale of public concern and the significance of these issues it is remarkable how little psychological study there has been of processes that clearly have a crucial psychological core to them.

In his helpful opening chapter to the present volume Timothy O'Riordan discusses the variety of perspectives that have to be taken into account when studying the perception and management of risk, thereby indicating the multidisciplinary perspective that needs to be embraced if risks are to be effectively managed. Many psychologists are uncomfortable in such a sea of viewpoints and this is probably one reason why there have been so few of them to tackle these important topics.

The dearth of research serves to enhance the significance of the studies that have been published in the *Journal of Environmental Psychology*. Bringing them together in one volume therefore provides direct and convenient access to the original studies that are laying the basis for a growing area of significant research. In his review of these papers Timothy O'Riordan does point to many strengths in these ground-breaking publications, but he also clearly identifies a number of topics that are crying out for careful psychological study.

Perhaps paradoxically for issues of such current topicality, the areas of psychological study that are opened up by the consideration of papers in the present volume are also central to many questions asked in the heartland of academic psychology; the most appropriate ways to categorize cognitions, the impact of context on perception, ways of summarizing and resolving differing conceptual systems. The present volume will thus be of value to cognitive and social psychologists who wish to inject major real world issues into what might otherwise be arid debates.

All scholarly study unfolds from the challenge and counter challenge of academic discourse, a process which takes time and does not respond well to pressures outside of the immediate academic concerns. The vast problems of environmental risk may often appear too urgent or complex to yield to empirical research. However, careful consideration of the studies in the present volume, read in conjunction with the helpful introductory chapters of its editor, will reveal that it is precisely because these issues are so significant that they must be subject to careful scientific examination. Bringing original studies together in one volume will be an important impetus to that process.

David Canter, Managing Editor
University of Liverpool

INTRODUCTION: RISK MANAGEMENT IN ITS SOCIAL AND POLITICAL CONTEXT

TIMOTHY O'RIORDAN

School of Environmental Sciences, University of East Anglia, U.K.

Risk Management as a Reflection of Social Transition

One can almost imagine the examination question in an environmental psychology degree course:

'Risk is a figment of the imagination' Discuss

It will be some time yet before the management of risk successfully unites the ideological divide between the natural and social sciences. For nearly a decade, analysts sought to make a distinction between 'expert' and 'perceived' risk as if the former revealed a true physical and statistical phenomenon while the latter was a matter of emotional irrationality (see Rowe, 1977, for an early statement). The whole purpose of the contemporary debate is to show that the two approaches are each in themselves only a partial explanation of the measurement and acceptability of risk. The immediate aim is to establish a cultural theory of risk that embraces both the deterministic and numerical scientific cultures with the qualitative, experiential and humanistic approaches associated with some of the social sciences.

A trans-scientific explanation of risk management is an arduous and often ill-rewarded task. Some continue to champion its cause because there is an element of economic rationality to be found. They argue that there is some point where the control of danger, of whatever manifestation, is not justified by the expenditures involved. As we shall see, that fulcrum shifts for a whole host of reasons, and it would be crazy to seek some comparable yardstick (see the chapter by O'Riordan *et al.* in the pages that follow). But the need for some kind of ordered and predictable economic–social risk–benefit framework stimulates a pragmatic and well funded research drive for further examination of this elusive goal.

Another reason for seeking greater reconciliation between various scientific approaches to the management of risk is that such an investigation throws light on the role of science itself in the tortuous process of policy-making. This in turn reveals the changing power base of expertise in a suspicious and temperamental society that searches for greater rights to know and demands to be able to control its own social and environmental health. Thus the contemporary risk debate encourages a shift in political perspective on individual and collective rights concerning what dangers should be tolerated and what should not. This is part of the reasoning behind the distrust of certain aspects of biotechnology, particularly where applied to genetically modified foods and animal welfare issues, notably in experimental research. For a fine assessment of this problem see Durrant (1992) and Roberts and Weale (1992).

The demand for a more 'accountable science' is also an issue in the cry for greater liability commitments to any carrier of danger or possible environmental damage, whether the owner of an oil tanker or the operator of a hazardous waste incinerator. There is a mood about that any possible danger must carry 'no fault' insurance and full compensation subject only to the arbitration of an independent tribunal. Similarly a case can be made for a community-instituted fall-back position that would carry a veto over any waste repository should it fail to meet the strictest operating standards. This point is touched on by Kemp (1992) in a fine assessment of empowerment politics over radioactive waste disposal, as well as by Clark *et al.* (1992). Both of these volumes contain a good coverage of the relevant theory plus original contributions as to how communities are struggling for improved rights over access to advance information and better control over what levels of danger (or safety) they are expected to face. These are demanding enough conditions: what is more revealing about this struggle for power is that some community groups want such safeguards to be prerequisites for the acceptance of an incinerator in their midst. This is a demand that can no longer readily be sidestepped.

This raises the theme of a fresh axis of communication between the vernacular science of popular gut feeling and the more sophisticated but less publicly accessible science of physical, chemical analysis and statistical prediction. At present such a relationship is a distinctly uneasy one; hence the suspicion with which any coventionally trained scientist would treat the hypothetical question set at the outset of this chapter. But the world of risk management is moving inexorably towards an imaginative and stimulating dialogue between these varying vocabularies, and this in itself is one of the most important outcomes of the contemporary risk debate. As we shall see, it has considerable implications for institutional design, concepts of risk–benefit assessment and the successful execution of risk communication.

Risk, therefore, is a cultural phenomenon. It is the outcome of a struggle over trust. It is the manifestation of the dispute over the freedom of action of an individual and a community in an increasingly undemocratic polity. It is the crisis over the loss of faith in scientific expertise especially where uncertainty is intrinsic to the analysis, in the face of a deep misgiving concerning the genuine independence of legislators in the face of technological advance or big money technological investment. For a good discussion see Ezrahi (1990), Jasanoff (1990) and Shrader-Freschette (1991). In all these ways risk management is a profoundly significant reflection of a modern, de-industrializing society in transition. It reflects the uncertain and oft interrupted search for more local empowerment; it captures a mood of pessimism over the political controllability of social change generally; and it drives into the heart of the world of science, a world which is ideologically fragmented and still the battleground of big egos and asymmetrically financed research programmes.

If one ever wants to sense a whiff of the current struggle within the scientific industry, read the introduction to the recent report by a Royal Society Study Group (1992), ably led by Sir Frederick Warner, entitled *Risk: Analysis Perception and Management*. The Royal Society executives who reviewed the final manuscript could not agree on whether the social science elements on perception and management should be allowed to appear in an unmodified form. In the end they acquiesced to the publication with the prefatory caveat that the report could not be issued as a whole, but only as six independent chapters attributed to those listed as authors. In short, the two cultures simply could not agree. Some of the allegedly 'non-scientific' statements and conclusions in the chapters on risk perception and management were regarded as too radical and unsupportable for the natural scientists to accept without considerable qualification. But at least the report was published in full, and this alone is a sign that the social science dimension in risk management for effective policy-making remains a necessary component in any comprehensive science research programme. That would not necessarily have been the case a decade or so ago. The changing social context of risk management is catalysing scientific reform.

Some Definitions

It follows that no definition of risk will ever be universally accepted. The already cited Royal Society report chose to define risk as

> a combination of the probability, or frequency, of occurrence of a defined hazard and the magnitude of the consequences of that hazard. (p. 4)

Risk management therefore becomes

> the process whereby decisions are made to accept a known or assessed risk and/or the implementation of actions to reduce the consequences or probability of occurrence. (p. 5)

Risk is therefore an amalgam of a danger, or hazard or events with adverse outcomes on health, well-being or safety, together with the likelihood of its happening. It follows that any definition of 'hazard' carries with it views as to what is good and bad about any set of predicted outcomes. This in turn introduces the concept of 'stakeholders' or individuals or groups with a 'stake' in any outcome—their values, how much they know, or are allowed to be informed, their capability to assimilate any such information or be helped to do so by science translators, their effective power to persuade, cause embarrassment or threaten sanctions, and their capacity for political mobilization. Similarly the notion of 'likelihood' creates a lively interest in whose guess is best, how it is arrived at, what authority it carries, and through which forms of legitimation it generates societal tolerance. Brian Wynne (1992) provides a provocative assessment of the failings of the 'normal' scientific method to assess risk, and hence the growing reliance on contested 'post-normal' scientific dispute, to arrive at an uneasy consensus over the determination of 'likelihood'.

Both risk and risk management are concepts and practices that cannot be dissociated from the institutional arrangements that create attitudes of mind, that encourage or discourage a genuine evaluative search for the most socially optimal solutions, that allow for stakeholder realignments and novel approaches to cost benefit accounting, and that enable political power broking to flourish.

It also follows from all this that risk and risk management are socially construed: they cannot be divorced from the personal, experiential and institutional settings within which risks are created and judged. The very processes through which risks are identified and managed are themselves objects of intense enquiry. This leads to the unsettling proposition that there is no single dimension of risk or risk management, no unique context and no comparable set of regulatory rules and practices. It was this relativistic and humanistic interpretation of the risk management process that gave the Royal Society senior echelons so much genuine concern. See Brown (1989) for a summary of this position as well as the oft-cited text by Douglas and Wildawsky (1982).

The observant reader will readily appreciate that only a few of the readings that follow cover such wide-ranging intellectual terrain. Social psychologists have made a vital contribution by applying decision theory techniques to the analysis of risk acceptance and avoidance. They have also established the principle that risk aversion is a function of both dread and a sense of lack of personal control over matters that could affect one's children and grandchildren without adequate redress. Yet psychologists have shown a tendency to be too tied to established frames of reference in these matters, leaving the all-important underlying motives, as well as the wider power-connected themes, to others to explore. Psychologists would be wise to expand their perspectives, and it is heartening to see that many are now doing so, as Pidgeon's (1992) chapter in the Royal Society report suggests.

Trends in Risk Management Research

Risk management research is essentially the product of the past 20 years of effort, with a special concentration of activity in the decade of the 1980s. The initial breakthrough was to demonstrate that there was no independently validated notion of risk, and that any actuarial engineering judgement was meaningless if separated from its social context. It was no good producing curves of frequency set against fatality, for the very context of fatality was a matter for interpretation. Someone dying of AIDS who was the victim of an institutional failure to screen blood in a transfusion could not be compared with a drug addict who contracted AIDS by refusing to eschew uncontaminated needles. As is pointed out in the chapters that follow, the Chernobyl disaster fundamentally transformed the social, political, technological and economic context of nuclear power. Though it must also be said that nuclear power is deeply associated with a fear of nuclear weapons, and a dread of uncontrollable contamination. These points are picked up by van der Pligt and by Eiser and his associates in the chapters that follow.

This led in turn to a second development, namely a grudging acceptance of the incorporation of social psychological perspectives in risk management, and a flurry of articles in journals such as the *Journal of Social Psychology* and *Risk Analysis* all of which are well documented in the Royal Society report. Of especial note was the empirical work of Slovic, Fischhoff and their colleagues at Decision Research in Portland, Oregon. See Slovic (1987) and Fischhoff *et al.* (1981) for a comprehensive statement of an oft-repeated position. These researchers sought to show that there was a fundamental characterization of danger in terms of what was familiar or dreaded, individualistic or catastrophic, voluntary or imposed, and that these dimensions resulted in great variations in the acceptance of risk in a wary society. Furthermore they tried to reveal a vocabulary of dread that actually froze in people's minds a fundamental fear and distrust of certain technologies and hazards. This locking in of mental constructs created a cognitive barrier and even a socialized resistance to communication about risk to the extent that any attempt to 'clarify' probability or to provide 'comparison' with better understood phenomena was doomed to failure. Indeed, such is the deeply rooted nature of cultural association between certain toxic substances and dread, any attempts to mollify the public with bribes, compensation payments or job creation schemes are treated with derision and become politically counter productive. Slovic and his colleagues (1992) show how this is the case in the endless dispute over the siting of a high level nuclear waste repository in the U.S.A.

Nevertheless this psychometric paradigm retains its powerful intellectual grip on social psychologists. This approach sees the characteristics of risks as fundamental to their acceptance. Yet various attempts to replicate the original Slovic study do not show the universality of a twin axis of 'dread' and 'catastrophe'. A comparison of such studies shows that various risks were rated progressively lower from wealthy and open culture countries to the poorer and more dictatorial regions. In addition Norwegian, Hungarian and Polish studies have shown a greater willingness to accept high tech risks compared with the U.S.A. This suggests that there are many more dimensions than a simple two axis framework would provide, and that the individual risk perceiver cannot be treated as mere statistical background. In the chapter by Cvetkovich and Earle that follows, there is a useful review of the early criticisms of the psychometric approach. Of note is the comment that it provides too facile a perspective on the socio-cultural reactions to various technological hazards, by no means resolving the disputes that rage within the conscience of any individual, let alone a community as a whole.

The third development was the incorporation of risk management into a wider cultural theory of blame, hierarchy, authority and denial. This was led by Aaron Wildawsky and Mary Douglas in their significant early work, *Risk and Culture* (Douglas & Wildawsky, 1982), but has subsequently been extensively developed by Shrader-Freschette (1991), by Michael Thompson *et al.* (1990), and, most recently, by Mary Douglas herself (1992) and by Krimsky and Golding (1992). Here the emphasis rests on the world view of various sectors of society depending on whether they accept a fatalist, authoritarian, participatory or interventionist approach to change and regulation. Risk management becomes not just a product of institutional culture, but of the very structures that both create and regulate hazard in modern society.

This suggests, but does not prove, that risk is not only culturally determined, but an artefact that enables people to defend or protect against activities or procedures that support or threaten their lifestyles or cultural values. Thus for some, nuclear power is a saviour technology while for others it represents a deeply resented faustian bargain. For those who believe in the relentless march of scientific and technological advance and who accept the avuncular role of paternalistic regulation, novel foods or reformulated pesticides are a boon. To others, whose cultural values embrace naturalness and a desire to control the non-chemical intrusions in their lives and in those of their descendants, the spectre of novel foods or of new agrochemicals is an insult and persistent source of anguish. In such contexts, risk management, no matter how sophisticated, can never satisfy.

This discarded transition has in turn drawn in different social science disciplines. At the outset, the economists were interested in the risk–benefit tradeoff. We shall see that this has enlarged to a more profound debate over what exactly constitutes acceptance as opposed to risk tolerability. Acceptance implies willingness on the basis of adequate knowledge. Tolerance is altogether a different term, suggesting acquiescence in the light of trust in the regulatory arrangements that control safety and provide adequate safeguards.

The second phase emphasized the role of the social psychologist, many of whom appear in the pages that follow. In the next chapter, the main contributions of this research will be reviewed.

The third period triggered the cultural theorists, who were never absent from the scene, but whose time did not come until the economic–psychological waves left considerable voids in the quality of explanation. Now the political scientists are more clearly in the picture, as there is more and more interest in the operability and construction of risk management institutions. The lawyer is also evident, for risk regulation nowadays is an intensely legalistic affair with plenty of attention by the liability-insurance accountants. This is especially the case given the growing role of strict liability in the responsibility for compensation against damage. Various court cases indicate that compliance with regulation and reasonable attention to all known safeguards are not sufficient defences. Liability is now likely to rest on anyone who disposes of waste, or who owns a site on which waste leaks, irrespective of when or by whom it was placed there. Furthermore, banks and insurance companies are also liable if they have a direct financial stake in the operation. This is concentrating minds wonderfully.

Hopefully we are emerging into a more interdisciplinary age, first of all amongst the social scientists and subsequently across the social and natural sciences. The key lies in applied case studies of both tolerability and institutional redesign, particularly for the regulation of novel substances such as genetically modified organisms, and the siting of controversial waste facilities. In this endeavour, the continuing emphasis on *risk communication* as both an emancipating and empowering exercise, and on *risk tolerability* as a device for improving the troublesome risk–benefit calculus, are probably the most powerful driving forces.

Risk Communication and Risk Tolerability

Communication is a two-way process that only achieves its aim if it is based on a presumption of mutual respect and shared values (Kasperson & Stallen, 1992). The social science literature is filled with texts that show how social communication is distorted by manipulative use of language, information, jargon and abstract concepts that confer a sense of explicit or covert power and powerlessness. Studies of public acceptance of nuclear waste facilities, for instance, show clearly how the institutional and legal structure of the waste management agency helps to determine how it seeks to communicate risk and how far it is likely to be believed, and by whom. In a society that suffers from official secrecy, or where commercial confidentiality can always be paraded as a safeguard against excessive snooping, risk communication will always be problematic. The very asymmetry of knowledge and ignorance confers a sense of resented authority and begrudged manipulation of data and their interpretation.

Similarly, the so-called not-in-my-back-yard (NIMBY) groups are not just bloody-minded resisters. Kemp (1992) has shown how they act as a force to restore a sense of fairness in a debate that is otherwise illegitimate, and provide a vital source of education and trust for a community that feels friendless and powerless in the face of a decision-making machinery that holds too many aces. This has led in turn to a theory of *risk amplification* (Kasperson *et al.* 1988) through which risks are amplified from a state of tolerated acceptance to a battleground for the struggle of individual and collective rights over the regulation of danger. This is manifest in protest, loss of trust and regulatory challenge, which in turn leads to a call for more open membership of regulatory structures. The trend is to include representatives of community groups on risk review panels, to ensure that the debates of such panels are more open, and to hold hearings over proposed risk regulations so that a more contested level of risk tolerance is created.

The U.S. legislation that requires that all chemical companies publicize all their toxic releases in full, and earn the support of the local community before continuation (the so-called SARA Title III legislation), is the prime example of this approach. In the European Community the 'Seveso' directive also requires that all plants associated with major hazard must consult with their local population, and must produce a satisfactory emergency plan in the case of accident. In countries, such as the U.K., where commercial confidentiality and secrecy go hand in hand, such requirements take time to put in place. The interesting development is the willingness of enlightened companies to see this practice as good for public relations and corporate community care responsibility. After a decade of hiccoughs, the process of direct community liaison has finally taken off. But it is largely due to the change of attitude in senior industrial management, born out of a period of much conflict and suspicion.

The concept of tolerance is also problematic in that it cannot be separated from the cultural values and political biases that shape society's views of regulation itself. For example, the U.K. Health and Safety Executive (1992) has published a report on the tolerability of nuclear power stations in which the central argument is that there is an acceptable level of risk at one end of a scale, an unacceptable level at another end, and a 'tolerable' level in between. That tolerable level hinges on the practicability of reducing the risk by good management and through technological advance, by better dialogue and by improved debate over the relative merits of benefits versus costs of further safety measures.

Nevertheless the fundamental flaw in the tolerability approach is the failure to recognize the different perceptions of benefits as well as risks. For a sizeable minority, nuclear power and toxic waste incineration are disposable options for which alternative methods of treatment and power production apply. It simply cannot be assumed that the benefits of nuclear power are universally equated amongst the population. Here again is a theme for further extension of cultural theory. So tolerability itself is part and parcel of the regulatory culture and the social fabric that promotes risk and distorts the communication over its overall social value.

The significance of regulatory culture in determining the trust over and tolerability towards risk should be given prominence here. Throughout the world, risk regulatory agencies are shifting both their style and their composition. No longer are their advisory panels entirely composed of specialist scientists with authority and expertise as their trade marks. Increasingly such committees contain social psychologists versed in risk perception research, and members of public interest groups whose values reflect an element of suspicion and criticism over traditional regulatory practices. Similarly senior managers in many agencies are adopting a much more liberal stance on the dual mode of hazard tolerance, and recognizing that community involvement, even though slow and at times very frustrating for all concerned, is the right way forward.

Managing Risk: Theoretical Dimensions

Much of this introductory chapter has benefited from Chapters 5 and 6 of the recent Royal Society report. For a truly excellent bibliography as well as a fine overall assessment of risk perception and risk management, the reader is strongly advised to read that report in full.

Risk management is in part a reflection of a changing concept of guilt and blame. In law, any defendant is presumed innocent until the process of law (not always reliable) has proven guilt beyond reasonable doubt. For many classes of hazard, particularly those that are novel and liable to create involuntary and potentially society-wide danger, the burden of evidence has shifted from the prosecutor to the defendant. Thus it is now more likely that a creator of any such substance or process would have to prove beyond reasonable doubt that there would be no serious hazard, and not upon the would be victim to show harm. This shift in the burden of proof is becoming familiar in the application of the precautionary principle to environmental management in general.

The application of the precautionary principle in risk management emphasizes risk aversion in society, the cultural biases against technological innovation, the need for a 'buffer' against ignorance, and the requirement of appropriate insurance arrangements and compensation clauses in liability protection in the event of miscalculation. These are profoundly challenging concepts. To some they represent a major departure from the ascendency and triumphalism of expert science and management, the unstoppability of innovation and the opening up of almost any policy decision to public challenge. To others this is the very basis of full throated assessment of future options and decisions, made in an open and participatory way, with the deployment of market forces to ensure that the burden of risk falls fair and square on the risk creator. The net result is a more formal and structural management process that is both flexible and more finely tuned to uncertainty and adaptation. It is less catastrophe prone, less liable to lurch unsteadily from one crisis to another, and less subject to political leg biting when an accident or fault occurs, even if genuinely unexpected.

The Royal Society group takes this idea a few steps further by suggesting the clearer identification of who is to be blamed for a disaster, by what processes or institutional structures did they fail to do their job properly, and what was the culture of management that led to them acting the way they did. This is the kind of reasoning that led to the charge of corporate responsibility for the Zeebrugge ferry disaster, and to the culture of incompetence that helped the oil tanker *Braer* onto the Shetland rocks. The law is a little unclear on the topic of corporate culpability: improved case law on the matter of corporate 'risk-taking cultures' in the face of risk adverse or precautionary regulations would sharpen up the defence. One can understand why some in the Royal Society were so uneasy about this line of reasoning.

The third main point to arise from the Royal Society review is the integration of risk management with management generally. In the area of pollution control, this not only involves integrated 'all round' efficient reduction, but risk prevention at source. Similarly in the new world of environmental auditing the idea of life cycle analysis means that products are analysed for their complete environmental and risk burden and managed accordingly.

These are very early days. Environmental auditing is still a fledgling art form with precious little case experience. This would involve the complete review of a firm in terms of all its environmental and social consequences weighted according to the burdens imposed. The comparability of burdens, such as fugitive toxic emissions from an operating chemical plant *vis-à-vis* hazardous solid waste that results from a state of the art incinerator remains a matter for considerable refinement. But one can find integrating concepts such as contributions to loss of life or injury, or to ecosystem decay or to global warming that do provide useful yardsticks. Within this activity comparative risk assessment has its role, both in terms of providing a common language, and for revealing how management in the round can lead to overall efficiency and socio-environmental well-being.

The very momentum created by such often discussed topics as precaution, environmental taxation, trading of emissions and risks rights, and various formal audits is pushing risk regulation along in a variety of exacting if not fully anticipated ways. The kind of vocabulary now used by risk-culture analysts is beginning to create its own special currency. In those, it has to be said, environmental psychology of the kind outlined in the papers that follow has had a vital preliminary role. But the field is moving rapidly onwards. It would pay the psychologists to team up with lawyers, insurance economists, ecological economists and public administration theorists if they want to keep ahead of the onrushing pack.

References

Brown, J. (ed.) (1989). *Environmental Threats: Perception Analysis and Management.* London: Belhaven Press.

Clark, M., Smith, D. & Blowers, A. (Eds.) (1992). *Waste Location: Spatial Aspects of Waste Management, Hazards and Disposal.* London: Routledge.

Douglas, M. (1992). *Risk and Blame: Essays in Cultural Theory.* London: Routledge.

Douglas, M. & Wildawsky, A. (1982). *Risk and Culture: An Essay on the Selection of Technological and Environmental Dangers.* Berkeley: University of California Press.

Durrant, J. (Ed.) (1992). *Biotechnology in Public.* London: Science Museum.

Ezrahi, Y. (1990). *The Descent of Icarus: Science and the Transformation of Contemporary Democracy.* Cambridge MA: Harvard University Press.

Fischhoff, B., Lichtenstein, S., Slovic, P., Derby, S. L. & Keeney, R. L. (1981). *Acceptable Risk.* Cambridge: Cambridge University Press.

Health and Safety Executive (1992). *The Tolerability of Risk for Nuclear Power Generation.* London: Health and Safety Executive.

Jasanoff, S. (1990). *The Fifth Branch: Science Advisers as Policy Makers.* Cambridge, MA: Harvard University Press.

Kasperson, R. E., Renn, O., Slovic, P., Brown, H. S., Emel, S., Goble, R., Kasperson, J. X. & Ratick, S. (1988). The social amplification of risk: a conceptual framework. *Risk Analysis,* **8**, 177–187.

Kasperson, R. & Stallen, P. (Eds.) (1992). *Communicating Risks to the Public.* Dordrecht: Kluwer.

Kemp, R. (1992). *The Politics of Radioactive Waste Disposal.* Manchester: Manchester University Press.

Krimsky, S. & Goulding, D. (Eds.) (1992). *Social Theories of Risk.* New York: Praeger.

Pidgeon, N. (1992). Risk perception. In Royal Society Study Group, *Risk: Analysis, Perception, Management.* London: Royal Society, pp. 89–134.

Roberts, L. E. J. & Weale, A. (Eds.) (1992). *Innovation and Environmental Risk.* London: Belhaven Press.

Rowe, W. D. (1977). *An Anatomy of Risk.* New York: Wiley.

Royal Society Study Group (1992). *Risk: Analysis, Perception and Management.* London:
    The Royal Society.
Shrader-Freschette (1991). *Risk and Rationality: Philosophical Foundations for Populist
    Reforms.* Berkeley CA: University of California Press.
Slovic, P. (1987). Perception of Risk. *Science,* **236**, 280–285.
Slovic, P., Layman, M. & Flynn, J. A. (1992). Risk perception, trust and nuclear waste:
    lessons from Yucca Mountain. *Environment,* **33**(3), 6–11, 28–32.
Thompson, M., Ellis, R. & Wildawsky, A. (1990). *Cultural Theory.* Boulder Colorado: West-
    view Press.
Wynne, B. (1992), Misunderstood misunderstanding: the social basis of expert credibility.
    Public Understanding of Science, **1**(3), 271–294.

THE COVERAGE OF RISK PERCEPTION IN THE *JOURNAL OF ENVIRONMENTAL PSYCHOLOGY*

TIMOTHY O'RIORDAN

School of Environmental Sciences, University of East Anglia, U.K.

This collection of papers represents a fair cross-section of psychological research on risks as published by the *Journal of Environmental Psychology*. By curious coincidence the first issue of JEP, which is obviously geared to covering a wide array of subject matter of which risk perception studies play a small part, coincided with the inaugural publication of the US journal *Risk Analysis*. This is the mouthpiece of the Society of Risk Analysis, a body with a lively European chapter in addition to its well-subscribed American cousin. *Risk Analysis* covers both the quantitative and qualitative aspects of risk estimation in a fairly equal measure. More recently it has emphasized the toxicological discussions of the risk research world, though it still carries a number of good articles on the politics, ethics, psychology and economics of risk perception. During the 1980's *Risk Analysis* held sway in this area. But the *Journal of Social Psychology* attracted a goodly measure of prominent researchers, and published some notable papers. Despite its much wider psychological coverage, the JEP has more than held its own in the risk field, a reflection of the assiduous work of its various editors.

The papers that follow are selected on the basis of both merit and representativeness. The text is divided into two main sections. For a brief period, social psychologists were anxious to devise frameworks for classifying hazards to fit the early psychometric work of Paul Slovic and his colleagues on decision research in Oregon. This work looked as much at how scientific paradigms could be translated into the psychological literature as at the specific distinctiveness of various hazardous processes or activities. The bulk of the text naturally covers risk perception research, for this is the heartland of the social psychologists' contribution. In general the papers that follow are dominated by studies linked to the nuclear industry, because the 1980s saw the leak at Windscale, the commissioning of the Sizewell B pressurized nuclear reactor, the Chernobyl accident and the long-running debate over the siting of radioactive waste disposal facilities in the U.K. This was a period when the public love–hate over nuclear power reached an unmatched intensity. Even a powerful prime minister such as Mrs Thatcher could not entirely get to grips with the public disquiet. So the field was wide open for well-financed social psychological research to assist the authorities to understand better just why there was so much public opposition to this technology. Yet good work was also produced in other areas, such as traffic safety and children's playgrounds, as well as the siting of overhead power lines.

Placed in perspective, the papers in the volume reveal how significant is this strand of academic research. They collectively show how techniques in perception studies have become more sophisticated with the introduction of a variety of powerful statistical techniques and clever approaches to questionnaire design. They also reveal the power of the comparative study, not so much to reinforce a theory as to highlight the underlying cultural dimensions of risk perception studies. The JEP has always prided itself in promoting international research perspectives, as the Chernobyl sequence demonstrates in the papers that follow. Finally they show that attitudes and judgements are selected out of a mosaic of feelings and experiences as mediated by events and social networks.

Possibly not enough work has been done on this last topic, though admittedly it is difficult to do well. We have always known that individuals absorb their sense of values more from the peer groups with which they are associated or whom they admire. Now that such networks are becoming more fragmented with the emergence of a much more mobile society, and authority is substantially diffused with the ubiquitous appearance of 'expertise' on television and radio. So it is possibly appropriate that a series of comparative studies be undertaken to assess the varying effect of social influences on individual judgements about hazardous products and processes, compared to the more indirect effect of authoritative opinion.

As Canter and his colleagues point out in the introductory essay, the rise of more structural techniques of risk assessment has been associated with the growth of greater formality in planning and decision-taking procedures generally. The growth of the environmental assessment industry, the introduction of legislation that requires risk assessment for statutory regulatory authorization of products or factories, the emerging art forms of whole-plant environmental auditing and life-cycle analysis of products—all of these have placed much more credibility on psychological studies of risk tolerability. The danger is that such techniques are being abused in order to get rapid results. Environmental impact assessment is neither cheap nor easy if undertaken properly. Much of a good EIA requires original data and painstaking community interviews, often involving groups and not individuals. Rarely is this the case: all too often the attitudes of residents are assumed as are the toxicity studies of emission data. The trouble is that few clients are prepared to shoulder such a burden if they feel they can get away with a 'cook book' copy.

So another area for fertile work is the faithful display of comparative environmental burdens associated with life cycle analysis or whole-plant environmental audits. Environmental burdens are the various 'costs' associated with a particular process or activity or substance. Some are well understood, for example the acidification chemistry associated with coal burning. Others are far more ephemeral, such as the possible dangers of genetic manipulation. Yet there is merit in devising techniques for displaying how different burdens can be identified and valued by people, utilizing the comparative risk assessment techniques developed by social psychologists. Again this would require fairly arduous and expensive research techniques, possibly involving panels or focus groups and fairly extensive informative material. The actual methodology itself is clearly going to prove very demanding.

Classifying Hazard

Cvetkovich and Earle seek to classify hazards along a life cycle of three stages: the hazard's causes, its social and psychological characteristics, and the individual and aggregate response. They correctly argue that such an approach should assist hazard management generally by 'aiding intuitive judgements, increasing analytical cognition, and reducing or at least clarifying, scientific disputes'. They also believe that a life cycle classification should assist in the proper design of emergency procedures as 'knowledge of cognitive strategies... may increase understanding of denial or passivity in the face of life-threatening events'. They produce a table that relates the generation of a hazard to its individual and social reaction potential. They also serve to highlight the human error dimension of many hazards: technology is only as safe as it is handled. Thus they stress the weakness of such distinctions as technological vs. natural risk, or man-made vs. national hazard. They point out that psychologists can make a useful contribution by incorporating the processes through which individuals react through such mechanisms as somatic or physical reactions to stress, coping behaviour, disorganized functioning and expressive behaviour, and various subjective reactions to stress.

As pointed out in the introductory chapter, one is left with a slight feeling of dissatisfaction over attempts to classify hazards, even by such clearly defined devices as catastrophic and society-wide; cataclysmal and individual specific; and 'daily hassle'. Yet the human being has a deep penchant for classification generally, so doubtless this attempt will not be the last. The richness of the socio-cultural context surrounding any hazard makes any standardized classification somewhat misleading.

Sally Macgill looks at an interesting example of legitimating science. Following the reported leak of radioactive material at Britain's only nuclear waste fuel reprocessing plant at Windscale on the north-western coast of England, the government appointed a distinguished epidemiologist, Sir Douglas Black, to report on the evidence of the connection between leukaemia and radioactivity and, in effect, reassure the public. In the event Sir Douglas was less than fully reassuring. This ambivalence enabled the social-cultural dimensions of risk perception to be fully aired. Those that supported expert judgement were satisfied, those that genuinely believed in the leukaemia connection because of the steady upgrading in radioactive exposure standards, used Black as an additional weapon to get the standards tightened up even more to the point where the reprocessing plant might become inoperable. Macgill concludes

> An important implication is that the establishment of trust between, on the one hand, concerned populations and, on the other, institutions in a position to offer reassurance, cannot be secured by a one-off initiative by externally appointed specialists offering expert reassurance.

Indeed a report such as the Black study does more to highlight the inadequacies of scientific knowledge and statistical techniques, reinforcing scepticism and suspicion among those who are convinced that any level of enhanced radioactivity is thoroughly dangerous, or that the nuclear industry in general cannot be trusted.

One of the great curiosities of the risk phenomenon has been the general passivity of those who live in radon-rich houses. In comparison with the Windscale example, there is far greater statistical likelihood of illness, though studies of lung cancer in occupants of radon-prone housing do not show any statistical significance. In the Svenson and Fischhoff study an interesting finding is that quite a few people simply do not want to know what the radon levels are in their houses. The authors describe a simple decision matrix of action and inaction each with varying money and stress consequences. This has a bearing on how the regulatory authorities play the game: too much action may prompt calls for compensation, too little may increase a sense of anxiety and impatience.

Risk Perceptions

Van der Pligt's study of salience and anxiety in attitudes to nuclear power seeks to show that there is an inner rationality to all perceptions in risk. His study reveals major differences in world view between pro and anti-nuclear activists with the anti-types placing such value on peace of mind and health of the community as a whole. This is a half way house to the cultural theory outlined in the introductory chapter: this study showed that underlying attitudes to technological advance, industrial modernization and growth in employment generally separated the nuclear supporters from the detractors. He concludes 'the fact that public attitudes are relatively stable *and* embedded in a wider context of values suggests that large scale attitude conversion may be more difficult than often assumed'.

The Chernobyl accident profoundly affected the perception of the scale and longevity of nuclear hazard. Even today we have little sense of what is going on around the plant itself where an area of over 50 km radius is severely contaminated. Doubtless there are Russian statistics on the subsequent health effects on the local populations, but little has yet surfaced in the West. It would be extraordinary if there were not quite a considerable public health burden in that most unfortunate of regions. The spread of radionuclides throughout Europe also created a sense of ubiquitousness and reality to a nuclear accident where whole hosts of innocent victims from sheep to reindeer to their owners and consumers could be identified. Compensation was not paid across the Russian border: national governments had to stump up for the mistakes of others.

Nevertheless it is doubtful if a country in energy need would seriously have banned the nuclear option even in the wake of Chernobyl. Admittedly the Italians did, but they would almost certainly have done so in any case. Elsewhere, the fortunes of the nuclear industry were far more influenced by that most mundane of influences, namely the money markets. The privatization of the electricity industry in the UK finally forced the true economics of decommissioning to be fully exposed. Two public inquiries in the U.K., one at Sizewell in Suffolk and the other at Hinkley in Somerset, had failed to get this crucial information. So much for openness and the legitimacy of public inquiries. Nowadays the nuclear industry remains in the doldrums awaiting favourable renaissance as the global warming debate hots up and the strictures of the UN Framework Convention on Climate Change force governments into non CO_2 creating energy generation.

The decommissioning costs will always be problematical for the industry, despite a favourable billing in the face of climate change. The main reason is the public antipathy to the encasement of retired nuclear stations and the demand for the highest standards of disposal of spent irradiated parts. Frankly no-one can be sure how a public in 100 years time will react to the technology of decommissioning, hence the difficulty in assigning costs. No wonder this element of the nuclear cycle remains in state hands.

The study initiated by Richard Eiser and his colleagues at the University of Exeter sought to discover how undergraduates responded to the Chernobyl accident on the basis of various psychological devices such as avoidance, dissonance and discrimination. In general they found that respondents assimilated their reaction by accommodating to pre-existing schemata thereby lessening the absolute size of any ultimate judgement. Thus the main factors influencing response were related to the degree of attention and anxiety that the students gave to the accident, set in the context of their fundamental pro or anti nuclear attitudes. Hence, it is not surprising, as Renn reports, that public opposition to nuclear power rose and subsequently fell back to roughly pre Chernobyl levels over a period of three to five years in most countries. This is largely because people experience a phase of cognitive dissonance when confronted with a possible but tangible health risk. They overcome this by denial or short-term frantic opposition, or by greater discrimination of the costs and benefits. In this transition, the press played a relatively laudable role, avoiding scare stories for the most part, and seeking as responsible coverage of the science as was possible, given the enormity of the event and the great array of conflicting evidence.

Another feature of the post Chernobyl period was to give more credence to the anti nuclear lobby, which gained a degree of popular support for its role in providing alternative perspectives. If nothing else, this kept the regulatory authorities on their toes, and considerably sharpened up emergency examination procedures, aimed at worse than the design-based accident. That goes against the technological grain, but it cannot be avoided these days.

Reaction to power lines depends in part on whether power lines exist in the locality already, and in part on the rise and fall of scientific and quasi-scientific prognoses on the health effects of high voltage transmission. Furby and his associates examine the structure of opposition, based mainly on a model of feelings about the power lines themselves and the subsequent siting process, as well as the pattern of conflict dynamics in the community. For example, as farmers become poorer, so they resist the minimal payments offered by the utilities for right of way rental of their land. Similarly, open landscapes are a source of much enjoyment. Power line siting usually seeks to follow the lines of existing routes to avoid the inevitable planning conflicts. But twinning high voltage transmission lines raises the spectre of asthma and other respiratory ailments. Furby et al. suggest that such is the power of anger or anxiety, that the very thought of a powerline in the vicinity produces symptoms similar to those claimed by people living in the shadow of the wires.

As is often the case with the geography of risk these days, there is an equity issue here as well. The dangerous or threatening facilities tend to be located where the populations are poor, ill-educated, or transient. Illegal immigrants are special targets for noxious and unwanted facilities. Articulate communities try to play the empowerment game. The weak, disorganized and inarticulate may lose out as a result. Articulate communities are usually involved in the planning process at the outset, the less articulate are not: levels of compensation also differ greatly. The politics of siting deserve as much attention as the psychology.

The study by Gärling and his colleagues on parental concern over children's traffic safety shows that the concern over safety in general, and the perceived traffic intrusion in the streets, showed up more as distinguishing variables than the actual danger as recorded in accidents. Also non-parents and parents showed remarkable similarity over their interpretation of possible road accidents to children. Where they differed was their assumption of cause. As might be expected non-parents were more likely to blame the child over a wider age range than were parents, who tended to see younger children as much more a factor in traffic accidents than older children. Also not surprisingly, parents saw crime as a much greater source of danger in the modern urban street than car generated accidents.

Baird's analysis of the perception of risks over a long time period reveal the well known outcome that people, for the most part, cannot see trends for more than 10–20 years. Beyond 20 years there is virtually no purpose in prediction for so much is unpredictable. Attitudes to long-term risks are bound up more with notions of progress and social–political management in the face of calamity than to the risks themselves. This is a general theme throughout this reader.

The media often come out badly in risk studies. Journalists are not specialists, and they operate in a culture that emphasizes newsworthiness rather than strict accuracy or balanced judgement. To be fair, too, most are at the mercy of their subeditors and editorial policy generally. This is rarely sympathetic to a sense of measured judgement over complex news stories where the science is convoluted and often indeterminate. So it is hardly surprising that Spears and his friends show that local newspaper coverage of siting controversies are usually hostile to the allegedly risky facility threatening its presence. Whether this opposition influences local opinion very much is a moot point. It has long been established that people select what they want from the media to reinforce pre-existing prejudices. It is unlikely that the influence of the local press is that great, but it is used as a convenient scapegoat by those whose proposals are fiercely opposed.

We looked at perceptions of risks over long time periods. Brown and her associates tried this from another angle, namely how children's drawings of nuclear stations differed pre and post Chernobyl. As might be expected, children do not draw the science, but their imagined world of the technology. She also found that children tended to depict more threatening associated features on the post Chernobyl diagrams, such as high security fences, warning and danger signs and police. But it is equally likely that they were portraying familiar objects, like a neighbouring factory, with various symbolic features attached. There is much that can be gleaned from drawings, but it requires sophisticated techniques to do this well, and it is always possible to draw too extravagant conclusions from quite innocuous sketches.

The so-called risk–benefit calculus of where a marginal investment in safety is justified by the expenditure is shown by O'Riordan and his colleagues as a matter of guesswork rather than analysis. The more controversial the safety issue, and the greater the uncertainty over the level of danger, the more sloppy the calculation. It is effectively a matter of engineering judgement based on what is technically possible. Cost only has a modest bearing. Where there is great public anxiety, the money for appropriate safety measures will be found even if the degree of safety improvement cannot be justified on analytical grounds. In effect, where there is technical or managerial scope for improvement, the job will be done, if for no other reason than it would be difficult to justify politically if the investment was not made. This raises the interesting question of the politics of precaution. Precaution is the process of planning from the worse than worst case just to be doubly sure, when the stakes of failure are so very high. The more informed and involved the public the higher the stakes, especially as technological advance proceeds. The very successes of high technology push the risk–benefit calculus almost to ludicrous margins. Such is the paradox of risk management in the modern age.

EDITORIAL: PSYCHOLOGICAL ASPECTS OF ENVIRONMENTAL RISKS

In our opening editorial to the first issue of the *Journal of Environmental Psychology* (Canter and Craik, 1981) we emphasized that the boundaries of our field should never be too firmly drawn or rigidly defended. Yet even we have been surprised and delighted to learn of the number of new concerns and fresh perspectives which continue to enliven the psychological study of the transactions between people and their physical surroundings. Over the past few issues, we have published articles on homesickness, restricted environmental stimulation techniques (REST), and the significance of the attic and cellar. No one can doubt that they all contribute directly to our growing understanding of person-environment relations, but who could have forseen them in 1980 when our *Journal* was conceived?

In part this divergence reflects the growing international structure of environmental psychology. We noted in the 1984 editorial the impressive number of international conferences and associations which now characterize the field. This development is reflected in our readership, which covers every major country of the globe from Australia to Yugoslavia, with China, Holland, India, Poland, Venezuela and many other countries in between. In particular, we have been pleased to see the growing number of high quality submissions from Continental Europe and have been happy to reflect this trend in the articles published in the *Journal*. We discern that environmental psychology in those countries does manifest distinctive styles and sets of concerns, which serve further to enrich our field of inquiry,

The perception of natural hazards is a research issue familiar to environmental psychologists. However, psychological aspects of technological risk and hazard is one of the newly emerging topics, hardly known five years ago but quickly gaining worldwide attention and pertinence, as recent events around the globe testify. The risk of hazards associated with our physical surroundings, especially those environmental components that are products of human activity and industry, have become increasingly important issues of public policy and public opinion. This field of investigation is such a burgeoning and important area of psychological research that we have made it the subject of the *Journal's* first special issue.

Psychologists and other social scientists have come relatively late to the study of environmental risk. The progenitors of risk research can be found in engineering reliability studies, actuarial analyses, and decision theories. O'Riordan (1983) has offered one account of why risk has become such a potent environmental issue. He argues that risk has moral and political dimensions that have to do with choices and social justice. Three particular characteristics give the study of risk its special qualities—the exercise of power in the distribution of risk, the rise of counter-establishment science, and environmental martyrdom. He concludes that the political climate in many industrial nations is responding to increasing public anxieties and disquiet about the future. More especially, doubts about the social benefits of new technologies and concern with the competence of regulators and the managing institutions are being expressed by some segments of the citizenry, coupled with frustrations in not being able to influence events.

Risk research attracts many disciplines beyond psychology, including statistics, epidemiology, biochemistry, engineering, medicine, physics, philosophy, economics and political science (Royal Society Study Group, 1983). As a consequence the many approaches, methods, conceptualizations, definitions and empirical studies present a bewildering array to a newcomer to this area of research. In 1981, the multi-disciplinary Society for Risk Analysis inaugurated an international journal, *Risk Analysis* (Cumming, 1981), as a common forum for the full range of approaches to the topic.

Within psychology, in an effort to focus on the perception and management of risk, a symposium was convened at the London conference of the British Psychological Society in 1982. The largely British-based research presented there resulted in lively debate and healthy disagreements which drew attention to the status and methods of risk analysis. A session of the American Psychological Association meetings held in Toronto the following year was devoted to environmental risk and engendered similar debate and reflection.

This special issure of the *Journal* represents the extension of some of the papers presented at these two conferences as well as additional articles by established researchers in the area of environmental risk perception. Some of our contributors also took part in the NATO Advanced Study Institute on technology assessment, risk analysis and environmental impact assessment, held in the French Alps in 1983 (Covello *et al.*, 1984). Thus, the articles in this special issue indicate the international interest in risk perception, its theoretical and applied implications as well as recent empirical investigations.

In modern industrial nations, an on-going policy discourse regarding technologies continues within agencies, legislative bodies and the mass media, and informal discussion among citizens. In complex ways this discourse seems to guide societal decision processes that lead to various environmental outcomes. The planning specialities of technology assessment, risk analysis and environmental impact assessment seek to inform the decision process, while diverse mechanisms more or less adequately monitor the environmental outcomes (whether they be ordinary functioning, malfunctioning, or disaster) (Craik, 1984). Collectively, the contributions to this special issue of the *Journal* illustrate persuasively the relevance of psychological research for advancing our understanding of each and every phase of this clearly imperfect, quite controversial, and still rather mysterious process.

Cvetkovich and Earle appraise systems for the classification of technological hazards. They offer an overview of the mutually productive task of using inquiry about risk to broaden the scope of environmental psychology but also to identify the contributions that environmental psychology can make to the management of risk—a theme also developed by Svenson and Fischhoff. Risk research is not simply a vehicle for environmental psychology; through its concepts and methods, our field has a tangible part to play in the assessment and management of environmental risk. For example, Cvetkovich and Earle propose that classificatory approaches can provide useful tools for management by (a) simplifying the structure of hazard and making relationships more apparent, (b) helping to determine priorities for risk management and (c) generalizing research findings.

Svenson and Fischhoff argue that environmental psychology, by documenting stressors that affect individuals' health and well being, can help to focus attention upon and ameliorate issues of environmental risk. Furthermore, they point out that at least as powerful a contribution lies in describing the worldviews and perspectives of opposing parties in environmental disputes. By using modelling methods from decision analysis they identify, for both individual homeowners and the authorities, events and consequences of decisions about remedial actions to limit levels of radon gas in dwellings. In explicating the individual citizen's view of what steps to take as well as the regulating authorities' options in setting acceptable levels of radon, they offer a procedure for illuminating and perhaps easing environmental disputes and conflicts.

O'Riordan, Kemp and Purdue (1987) appraise another mechanism for dealing with societal conflict about risk—that of the public inquiry. They conclude that the concept of acceptable risk 'has no absolute meaning or operational significance . . . (it) should only be regarded as a management and regulating guide . . . its operational value can best be revealed through a mechanism for encouraging informed dialogue and public debate.' Specifically, they describe the character of a public inquiry into the siting of a pressurized water cooled nuclear reactor at Sizewell in Suffolk.

The notion of diagnosis is used by Svenson and Fischhoff in pinpointing and elucidating areas of dispute, i.e. the description of different parties' positions and perspectives can lead to identifying sources of disagreement. In an indirect way, Cvetkovich and Earle also discuss the curative and preventive aspects of risk analysis, i.e. by providing opportunities to study anticipatory problems as well as helping to mitigate effects to the physical environment. Canter and Powell examine curative and preventive measures in the context of the management of a chemical plant. Here the focus is on the behaviour of the managers. The research task is to identify those human factors which make up standard good practice, such as staff selection training, and accountability. These functions suggest criteria for evaluating good practice. The central issue is the actions individuals perform in relation to risks, rather than their perceptions of or attitudes toward those risks.

Two contributions address the issue of public attitudes and perceptions of technological risks. Van der Pligt reviews public opinion on nuclear energy in Western Europe. identifying shifts in recent years. He argues that differences in the stance taken by individuals on nuclear energy is related to their more general values and beliefs about society and technology. Along with Cvetkovich and Earle, he discusses methodological issues in analyzing public perceptions of risk. The important issue of public communication and the role of the mass media in attitudes toward technological hazards, noted also by Cvetkovich and Earle, is represented by Wober and Gunter's empirical study of possible relationships between television viewing and perception of hazards.

The present contributions document the wide variety of psychological issues generated by an examination of the societal management of environmental risk. They also illustrate at least some of the diverse ways in which psychological concepts, findings and expertise might usefully enter into the on-going societal decision processes. Finally, they establish the importance of environmental risk perception and management as a central new research topic for environmental psychology.

This *Journal* will continue to welcome conceptual analyses, research reports and reviews dealing with this important set of issues.

DAVID CANTER AND KENNETH H. CRAIK
WITH JENNIFER BROWN*

* Jennifer Brown, the *Journal's* Book Review Co-editor, was the invited Editor for this special issue on the psychological aspects of environmental risk and acknowledges the contribution of Kenneth Craik in initiating the American papers and David Canter for his overall guidance.

References

Canter, D. V. and Craik, K. H. (1981). Environmental psychology. *Journal of Environmental Psychology*, **1**, 1–11.

Covello, V. T., Mumpower, G., Stallen, P. J. and Uppuluri, R. (eds) (1984). *Technology Assessment, Environmental Impact Assessment and Risk Analysis: Contributions From the Psychological and Decision Sciences.* New York: Springer-Verlag.

Craik, K. H. (1984). Psychological perspectives on technology as societal option, source of hazard and generator of environmental impacts. In V. T. Covello, G. Mumpower, P. J. Stallen and R. Uppuluri (eds), *Technology Assessment, Environment Impact Assessment and Risk Analysis: Contributions From the Psychological and Decision Sciences,* 211–236. New York: Springer-Verlag.

Cumming, R. B. (1981). Is risk assessment a science? *Risk Analysis*, **1**, 1–3.

O'Riordan, T. (1983). The cognitive and political dimensions of risk analysis. *Journal of Environmental Psychology*, **3**, 345–354.

O'Riordan, T., Kemp, R. and Purdue, H. M. (1987). *Sizewell B: An Anatomy of the Inquiry.* Basingstoke: Macmillan.

Royal Society Study Group (1983). *Risk Assessment.* London: Royal Society.

CLASSIFYING HAZARDOUS EVENTS

GEORGE CVETKOVICH

Western Washington University, Bellingham, Washington, U.S.A.

and TIMOTHY C. EARLE

Battelle Human Affairs Research Centers, Seattle, Washington, U.S.A.

Abstract

Classification is an information management technique that helps achieve the goals of scientific and applied research by simplifying and ordering complex and varied observations. This paper is intended as a guide to the use and application of classifications of hazardous events. Proposed classifications are reviewed and evaluated in terms of their purposes, methods, supportive evidence and the scientific and applied goals of classification. The review is structured according to the metaphor of the natural life history of a hazard that distinguishes three stages of human–environment interaction: (a) hazard causes, (b) physical and psychosocial characteristics of hazards, and (c) individual and aggregate responses to hazards. The life history perspective has the advantage of being multidimensional and transactional. The structure highlights the relationship between proposed classifications, identifies needed areas of conceptual and empirical development, and provides a general guideline for the selection and use of classifications for hazard management.

Introduction

Hazards are a special kind of environmental event that pose threats to humans and to the things that humans value (Hohenemser *et al.*, 1983*a*). They represent the potential occurrence of extreme conditions of the natural environment or the misfunctioning of the human-built technological environment. There has been a decided emphasis within environmental psychology on adaptation to routine and ordinary environments. The study of hazards is important in order to broaden the scope of environmental psychology and because it provides the opportunity to study efforts to anticipate problems, to shape and mitigate the effects of the physical environment, and to adapt to extreme conditions. Our concern is not only with what the study of hazards can do for environmental psychology, however, but is also with what environmental psychology can contribute to the improvement of hazard management.

The major purpose of this paper is to review proposed classifications of hazardous events. The perspective taken in the present paper is that classifications are useful tools for the management of scientific and practical information. Indeed, recent years have seen an increasing number of efforts to move beyond purely descriptive accounts of hazards, to develop more general comparisons and to classify events according to identified general characteristics precisely because of manifest information management advantages. For the most part classification efforts have remained separate, often exploring different subsets of hazardous events, using different methods for constructing classification schemes and generally progressing with little

Requests for reprints should be sent to George Cvetkovich, Department of Psychology, Western Washington University, Bellingham, Washington 98225, U.S.A.

recognition of the work done by others. The information management perspective, which is further developed in the next section of this paper, suggests that a next important step in the conceptualization of hazards is the integration of the diverse proposed classifications into a single framework. As the present review of the literature shows, such an integrative framework should be comprehensive and multidimensional in nature so as to reflect the diversity and complexity of numerous hazards. It should also reflect the transactional nature of the relationship between humans and the environment. The transactional nature of hazards is generally accepted by classifiers. It is assumed that physical characteristics of a hazard influence risk judgments and that, in turn, images of the hazard's physical properties are related to both individual and group hazard responses, which may include changing the physical environment. Because classifiers have each dealt with a circumscribed aspect of the total domain of hazards, however, conceptualizations and empirical research specifying how these interactions take place, how they change over time, and how they are influenced by contextual factors are lacking.

The overall scheme for integrating existing classifications proposed in this paper is both multidimensional and transactional in character and is based on the metaphor of the life history of a hazard. Three stages of the human–environment interaction occurring during a hazard are distinguished: (a) the hazard's causes; (b) the physical and psychosocial characteristics of the hazard; and (c) individual and aggregate responses to the hazard. As it happens hazard classifiers have restricted their efforts to one stage. Thus the meta-classification based on the life history metaphor highlights the relationship of the classifications to each other and points to needed areas of conceptual and empirical development. The major portion of this paper examines the purposes, methods and supportive evidence of each proposed classification within the life history framework. The classifications in each area are evaluated against the scientific and applied uses of classification, and the adequacy of supportive evidence. The problems encountered in the classification of events highlight the needs and inadequacies of hazard research in general. It is intended that this discussion serve as a guide to the use and application of the reviewed typologies. Suggestions concerning future directions for the classification of hazards and for hazard studies are made in the concluding section.

Why Classify Hazards? An Information Management Perspective

Classification, the process of identifying differences and similarities among groups of objects, entities, behaviors or other phenomena of interest, is an important aspect of science, as well as many other human activities. Much of the research of environmental psychologists entails the use of classification. In the effort to understand human–environment relationships, environmental psychologists have often found it useful to classify environments. Implicit or explicit classification schemes underlie most decisions concerning research designs, measurement and identification of the boundaries of areas of study. For example, a distinction is often implicitly made between the natural and the human-made environment (as illustrated in the organization of many environmental psychology textbooks; e.g. Fisher *et al.*, 1984). Examples of explicit classification include Stokols' (1978) distinction between primary and secondary environments in terms of affective meaning and commitment to a setting; Willems' (1977), Barker's (1968) and other ecological psychologists' distinction of units of the environment in terms of behavioral settings, and Moos' development of scales to assess the social climate in industrial, residential, academic, military and treatment settings (Moos and Gerst, 1974; Moos 1976). Beyond academic efforts, classification plays a pivotal role in the development of practical strategies for controlling the environment. For example, most discussions of hazard regulation policies include the classification of threatening events, activities and technologies (e.g. Rowe, 1977; Lowrance, 1976).

While the development of classifications can be recognized as an integral part of environmental psychology, other scientific disciplines and practical efforts to control the environment, two different general viewpoints must be distinguished. Much of the scientific work to develop explicit classifications has been aimed at devising systems that accurately reflect categories of phenomena (taxa of animals or plants for example) as they exist in nature. The assumption underlying such systems is that if it is the purpose of science to discover the true nature of things then it is the purpose of correct classification to describe objects in such a way that their 'true' relationships are displayed.

Sokal (1974, p. 116) has outlined three difficulties with this 'essentialist' orientation to classification. (1) The axioms of essentialist classification give rise to observation of properties that are inevitable consequences of the axioms, thus the classifications tend to be circular. (2) Taxonomists often develop classification systems consisting of discrete, mutually exclusive clusters of entities although natural phenomena may exist in the form of overlapping classes or classes with fuzzy boundaries, thus the structure of classification and nature do not match. (3) Most classification systems can not match the diversity, complexity and magnitude of natural systems; folk taxonomies all over the world are restricted in the number of forms of animals and plants identified (Berlin et al., 1968). This range is found regardless of the richness of fauna and flora in the area in which the people live. This suggests a cognitive limit on the number of distinctions possible within a classification system. Another limitation is the difficulty of gathering information to assign entities to classes. There are an estimated 10 million kinds of organisms in the world. Only about 15% of these have been scientifically studied and described. Given current extinction trends it is doubtful that even 5% more of the organisms will be described before they become extinct (Raven et al., 1971). Thus, classification systems may fail to match the 'essence' of nature because the systems are incomplete by comparison.

The above examples come from the domain of biological science, but the same concerns apply to the classification of hazardous events. Hazardous events also occur in a wide diversity of different forms and often have blurred overlapping characteristics (as will be illustrated later in discussing the distinction between 'natural' and 'technological' hazards). These characteristics, along with the problems inherent in the essentialist approach, would seem to preclude the development of a single comprehensive classification scheme.

An alternative view is to define classification as 'a human decision, constrained by a bevy of facts, about how best to order nature' (Gould, 1974; p. 740). This 'constructivist' approach considers classification systems as aids to thinking and communication and assumes that there is no single generally best classification system. The quality of a classification system depends upon how well the functions of analysis and communication are performed. Thus whether one system is better than another depends upon the specific aims of its user.

The classification of hazards, or any entities, aids thinking and communication by achieving two important consequences (Sokal, 1974; Sneath and Sokal, 1973). One advantage of classification is that it provides a means of easily manipulating information. If entities are arranged in a system of related groups (hierarchies or other arrangements) relationships are more easily discovered and retrieved. Ease of retrieval is often used as a criterion for judging the adequacy of a classification scheme (Sokal, 1974). This function of classification is being increasingly recognized by those who are attempting to make the rapidly expanding literature on hazards easily accessible (Lind, 1984; Greenwood et al., 1979).

A second advantage of classification is that it achieves a simplification of the phenomena under consideration. By grouping entities one can avoid having to deal with a world of single cases and individual sets of characteristics. When an entity is classified into a group, its characteristics are subsumed by those of the group. In essence a classification system is a generalization about related entities. This advantage is particularly important to the study and management of hazards. There is a multitude of events that could cause harm to life and property. The simplification achieved through classification could make relationships more readily apparent and thus contribute to the development of scientific theories concerning human anticipation of and responses to hazards (Slovic et al., 1983; Kreps, 1982) and the identification of 'generic' responses that can be applied to related hazards (Perry, 1983a).

In addition to easing information use and providing a simplified structure, a classification system of hazards can also be used to determine priorities for risk management and for helping to make management decisions about little known hazards. A classification system that allows ordered comparisons of hazards could be used to aid hazard managers in setting priorities and determining the amount of attention to give to each (Meyer and Solomon, 1984; Solomon et al., 1982; Hohenemser et al., 1983a).

Finally, a classification can be useful in the process of generalizing research findings. Some hazards occur very rarely or have never occurred before. If, on the basis of specified characteristics, a classification system permits these events to be grouped with other more frequently occurring (i.e. well studied) events, generalizations can be made. Perry et al. (1980), for example, have compared how nuclear attack is similar to better studied events such as natural hazards. On the basis of this comparison, predictions concerning compliance with evacuation efforts are made. Disagreements among scientists about generalizing laboratory findings, concerning hazards to real life circumstances, occur in part because such judgments are made intuitively (Hammond et al., 1984). A classification system could be a valuable technique for aiding intuitive judgments, increasing analytical cognition and reducing, or at least clarifying, scientific disputes.

The information management perspective provides several answers to the question of 'Why classify hazards?'. Classification systems improve communication, aid in the generalization of research findings, are useful in identifying lacunae in the scientific and practical body of knowledge about hazards, and, most importantly, can be valuable tools for improving analytical thinking and the practical management of hazards. Not every classification system performs these varied functions equally well.

One purpose of this review is to evaluate how well the classifications perform their intended functions.

A Life History Organization of Classifications

This paper reviews a number of major hazard classifications. The schemes each deal with different selected aspects of hazardous events and in many cases uses entirely different information bases. Given the complexity and variety of the phenomena classified and the divergent purposes of the researchers, it is not possible to develop an adequate single, comprehensive classification scheme. It is proposed that order can be imposed on these classification efforts by examining them relative to the natural history of a single generic hazardous event. A dynamic, transactional model of adaptation to hazardous events presented by Earle and Cvetkovich (1983) is adopted for this purpose. The applicable elements of the model are presented in Figure 1. This use of the model makes a distinction between three major stages in a hazard's life history: (a) hazard causes, (b) characteristics of hazards and (c) responses to hazards. The processes operating at each stage and the relevant hazard management strategies are indicated in the figure.

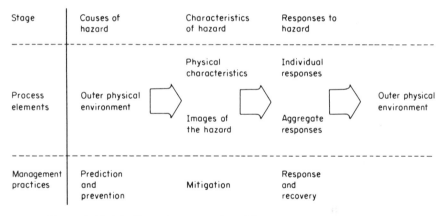

FIGURE 1. Model of hazard process (adapted from Earle and Cvetkovich, 1983).

Hazard causes
A hazard is an event originating in the environment that is threatening to human life and/or valued property. Traditionally, hazards have been distinguished by hazard managers and researchers in terms of two aspects of the physical environment; (1) hazards that are caused by events transpiring in the natural environment and (2) those caused by the built, technological environment. A more detailed discussion of the complexities of the distinction between natural and technological hazards is presented later. The purpose of research concerned with the causes of hazards is to enable their prediction and the reduction or elimination of negative consequences.

Characteristics of the hazard
Ideally a model of human–environment interaction should incorporate both situa-

tional demands and the mental activity engaged in by the individual who is attempting to adapt to a hazardous situation (Kaplan, 1983). This model includes these concepts by making a distinction between physical characteristics of a hazard and the subjective images of it. Examples of important physical aspects of a hazard are characteristics such as speed, extent and duration of possible destruction, presence of secondary impacts, degree of social preparedness, and population at risk. Variations in these and other physical characteristics define the situational demands of a hazard by influencing judgments about its nature and, importantly, determining the selection and implementation of plans and acts to deal with it (Perry, 1983a; Hohenemser et al., 1983a).

The image of an event as dangerous and a threat to self, loved ones and valued possessions includes a compound of likelihood and consequences generated in varying ways. The study of hazards commonly focuses on risks associated by individuals with specific activities and events. The model also is concerned with the overall evaluation of activities and events in terms of risks and benefits. Examples of the study of benefit judgment in conjunction with hazards include Vlek and Stallen (1981), and Gardner et al. (1982). Fischhoff et al. (1983), in an effort to define risk comprehensively, have provided a list of the psychological characteristics of risk decisions. The purpose of research on hazard characteristics is to understand the images that people hold. The practical outcome of this work is the improvement of communication and education programs.

Responses to hazards
There is a wide variety of actions that can occur in response to a hazard. Some responses are intra-psychic experiences, such as feelings of stress, others are individual and overt, and still others are aggregate responses, such as developing a community evacuation plan. Some responses attempt to eliminate directly the hazard and others are indirect attempts, such as leaving a hazardous area. Some responses attempt to reduce losses should a threatened event occur; examples include seeking shelter, purchasing insurance, developing disaster contingency plans and seeking information about the hazard. Other responses occur after a hazardous event has occurred. This class of behavior includes providing emergency services, attempts at recovery and rebuilding, and reducing the likelihood that the hazardous event will cause severe damage again.

The model adopts the perspective provided by the 'behavior–environment optimalization' concept (Stokols, 1977; Kaplan, 1983). Basic to this perspective is the notion that the human–environment interface involves the central theme of maximum fulfillment of needs and accomplishment of goals or plans. For the individual an incongruence between environmental demands and individual goals and needs produces stress and, in many cases, efforts to bring about a demand–need congruence. By definition hazards constitute a situation of actual or threatened incongruence.

Persons possessing a sense of control have been found to take more self-protecting steps in hazardous or obnoxious situations than those who believe that they have little control (Baum et al., 1981; Sims and Baumann, 1974; Glass and Singer, 1972). A single individual does not have complete control over a hazardous situation in the sense of being able to determine whether the event occurs or not. A feeling of being in control may be preserved, however, if the individual believes that adequate procedures and efforts to mitigate losses are in progress (Lacey, 1979;

Peterson and Seligman, 1983). Rothbaum *et al.* (1982) have proposed that people not only exert primary direct control, but may also gain a sense of secondary, cognitive 'control'. This secondary control is reached by finding meaning and value in uncontrollable events (Bulman and Wortman, 1977). One such cognitive strategy is to always expect the worst and thus to 'control' negative outcomes through anticipation. Knowledge about the use of such secondary cognitive strategies, and other 'palliative' coping (cf. Lazarus, 1966), may increase understanding of denial and passivity in the face of life threatening events such as hazards (Peterson and Seligman, 1983).

One important mode of adapting to a hazardous situation and of increasing control is the seeking of information. Weinstein (1979), studying adaptation to a new environmental health threat, found that theoretical formulations (e.g. Janis and Mann, 1977; Freedman and Sears, 1965; Frey and Wicklund, 1978) did not adequately encompass his results. Based on his study he offered a model consisting of two propositions: (a) individuals tend to select messages that agree with their points of view when messages disagree about whether a hazard exists and (b) information about a new issue is determined by general interest in a topic, whereas information-seeking about an old topic is a function of assessment of risk. As well as suggesting some of the complex of factors influencing information seeking, these conclusions also emphasize the active, striving nature of coping with an environmental hazard.

Stokols (1982) and Stokols and Shumaker (1982) has argued for the use of contextual analysis to study adaptation to the environment. As applied to hazard research, this approach would link images, plans and actions not only to the immediate circumstances surrounding a hazard, but also to the broader context of the individual's background such as socio-economic class, education and interests (Committee on Socioeconomic Effects of Earthquake Predictions, 1978; Mileti, *et al.*, 1981; Hirschman, 1981; Knott and Wildavsky, 1980; Miller and Barrington, 1981; Miller *et al.*, 1980; Tichenor *et al.*, 1976; Witt, 1976), current life circumstances, aspirations for the future, and values.

Environmental sociologists and others have rightfully persisted in making the point that social responses to hazards can not be reduced to the sum of individual cognitive processes and actions (Dunlap and Catton, 1983; Douglas and Wildavsky, 1982; Otway and Thomas, 1982). For this reason the model makes a distinction between individual and aggregate responses to hazards. Aggregate responses, such as the development of warning systems, disaster relief plans, evacuation programs and the identification of which hazards are worthy of concern are reflections of the operation of political and organizational forces (Nelkin, 1977). Public policies are not 'made' in the sense that individual decisions are, but are negotiated by competing groups and institutions (Linnerooth, 1983). The identification of the relationships between individual cognition, emotion and behavior and these larger forces is a task awaiting future research.

The practical outcome of research on individual and aggregate responses to hazards is the development of programs to deal with stress, feelings of lack of control and other negative psychological consequences of hazards, and the establishment of means to deal with hazards and to adopt and implement plans for recovering from disasters.

Summary

The proposed model makes a distinction between three logically ordered steps in the natural processes of a hazard: the hazard's cause, the characteristics of the hazard and responses to the hazard. In keeping with the model's transactional emphasis, hazard causes are identified both in terms of their physical characteristics which define situational demands and in terms of subjectively judged characteristics. Responses to hazards are divided into those of the individual and those of larger social aggregates in recognition of these two distinct levels of analysis. The remainder of this paper uses these distinctions as a framework for reviewing classifications of hazards.

Classification of Hazard Causes

Hazards can be conceived of as resulting from the interactions occurring among four general systems of adaptation and developmental progression (Riegel, 1975; 1976). These causes of hazards are (a) the inner-biological, (b) the individual-psychological, (c) the cultural-sociological and (d) the outer-physical. If the progressions of two dimensions are synchronized, positive growth occurs. If progressions are not synchronized, a crisis occurs which produces a negative outcome. For example, if a group's cultural-sociological life is synchronized with its physical environment, conservation of resources and continued survival result. If the relationship between the cultural-sociological and the outer-physical environment is asynchronous, devastation occurs. Thus, a hazard is an entity or process which has the potential of causing disequilibrium as one of its possible outcomes (Fiksel, 1981). A complete listing of outcomes generated by the interaction of the four dimensions is given in Table 1.

Table 1 permits the identification of the domains of interests of various scientific disciplines. The physical sciences have been traditionally interested in the content of cell P. The biological sciences have focused on cell D with a recent increasing interest in cell C. Medicine has focused primarily on cells A, B and, more recently, on cell E. Cell E is also a recent interest of individual psychology, added to its more traditional domain identified by cell F. Social psychology has as its domain

TABLE 1

Hazards generated by the interaction of four causal dimensions (adapted from Riegel, 1976)

Dimension generating hazard	Dimension experiencing reactive change			
	Inner-biological	Individual-psychological	Cultural-sociological	Outer-physical
Inner-biological	A. Infection	B. Illness	C. Epidemic	D. Deterioration
Individual-psychological	E. Disorder	F. Discordance	G. Dissidence	H. Destruction L. Devastation[a]
Cultural-sociological	I. Distortion	J. Exploitation	K. Competition	P. Chaos
Outer-physical	M. Annihilation	N. Catastrophe[a]	O. Disaster[a]	

[a] N O, H and L are the domains of interest of environmental psychology.

cells G, K, J and N. Three of these are shared with anthropology and sociology which also focus on the content of cells J, K and N, as well as C, G and O.

Of interest to environmental psychology are those hazards generated by the interaction of individual-psychological and cultural-sociological systems and the outer physical environment—cells N, O, H and L (i.e. hazards that are labeled catastrophe, disaster, destruction and devastation). These hazards are the major focus of this discussion. Events usually referred to as natural hazards can be seen simply as having effects generated by the outer physical environment on the individual-psychological and cultural-sociological systems (cells N and O). Technological hazards involve a more complex sequence. They are generated by the effects of the individual-psychological and/or cultural-sociological systems on the outer physical environment (cells K and L) which in turn affects the individual and cultural systems (cells N and O). The distinction between natural and technological hazards has a long standing history in the research literature (Craik, 1970) and it remains as a basic component of the profile of risk used by many contemporary classifications (cf. Rogers and Nehnevajsa, 1984). A brief review of its use demonstrates the problems of many simple descriptive schemes.

Natural vs. technological hazards
White and Hass (1975) define natural hazards as consisting of extreme geophysical events (hurricane, flood, earthquake, etc.). Hohenemser *et al.* (1982) add to geophysical events organisms that attack agricultural products and viruses that inflict humans. The National Research Council's Committee on Risk and Decision Making uses still another definition of natural hazards, including geophysical and 'certain physical conditions generated by nature (e.g., diseases of old age, genetic mutations, and diseases from natural ionizing radiation)' (National Research Council, 1982; p. 16).

The lack of agreement as to which hazards are natural and which technological suggests that this distinction may not be the most useful classification scheme. There have, nevertheless, been a number of attempts to preserve this distinction. Many of these have argued for the distinction on the basis of two factors: (a) the relative ease of prediction of disaster and (b) the extent and nature of human involvement.

Prediction of disasters. Hewitt and Burton (1971) propose that one reason the traditional distinction between natural hazards as geophysical events and technological hazards has so often been made is the relative ease of measurement and prediction of the former. This assumes that extreme geophysical events and their magnitudes and frequencies, if not their actual place and time of occurrence, are relatively well known whereas accumulated general knowledge of technological hazards is relatively new. This assumption is not entirely acceptable since knowledge about certain technological hazards (automobiles and earthen dams for example) has been accumulated for a number of years. While some extreme geophysical events may be relatively easy to measure and predict, other hazards labeled natural by some (e.g. viral infections and radiation effects) are not. Lawless (1982) argues that even among the so-called technological hazards there is a wide variety of predictability of likelihood and magnitude of effect. Some technological hazards involve repeated exposure of humans to risk (for example earthen dams, oil tanker operations, hazardous chemical rail cars). Relatively good data exist about the likelihood and magnitude of consequence of these events. Other technological hazards are unprecedented, but involve exposure of a relatively small number of individuals to risk (e.g. Toxic Shock Syndrome and tampons, a Skylab falling out of orbit). Still others are unprecedented and have a potential for catastrophic results. The collapse of a hotel skywalk, as occurred at the Kansas City Hyatt Regency, for example, while predictable on the basis of engineering data, had never happened before.

Nature of human involvement. Some researchers have argued that technological problems are more controversial than natural ones (e.g. Baum *et al.*, 1983*a*). According to this reasoning, technology is human-made and is designed to solve problems, not produce them. Thus, it is argued, humans are the central agents involved in technological hazards, but not in natural hazards. Rowe (1977) makes a four-fold distinction attempting to clarify the role of human responsibility for a hazard. 'Unavoidable natural disasters' are events such as the impact of large meteorites that are completely beyond the control of humans. 'Avoidable natural disasters' are ones over which humans have at least a limited ability to mitigate losses either by avoidance or control. Floods are one example of such events. 'Man-originated disasters' occur because of human activity and are distinguished into two groups: 'Man-triggered disasters' are not only created by human activity but also the result of it; 'Man-caused disasters' are events for which humans originate the conditions producing exposure to risk, but natural phenomena trigger the disaster. Examples of these two categories would be a disaster befalling people living down stream from a dam: if the dam failed because of an earthquake this would be a man-caused disaster; if the dam failed because of sabotage, poor construction or inadequate maintenance, it would be a man-triggered disaster.

Miller and Fowlkes (1984) have argued that the term 'man-made' disaster, not technological disaster, should be retained to convey the notion that these events are caused by human error that can be anticipated. They suggest that adoption of the perspective that disasters involving technology are caused by impersonal forces will result in three consequences: (a) the events will ultimately be judged to be unavoidable and an acceptable price of living in a technological society; (b) profitable, but unsafe management practices will be encouraged and (c) effective response to these disasters will be hampered by a failure to recognize the 'passions and political forces involved in their creation'.

While humans may not be the direct cause of natural disasters, they do often attempt to intervene in the natural environment both to reduce the likelihood that a hazard will become a disaster and to reduce negative consequences once a disaster has occurred. Sometimes these mitigating and palliative efforts, contrary to their intended purpose, may increase the risks involved or actually exacerbate losses. Moreover, failed efforts to mitigate natural hazards can certainly be blamed on human agents whether for reasons of incompetence or self interest. For example, within the last several years the emergency situations produced by the failure of public works departments to anticipate adequately the need for snow removal from city streets may have damaged the political careers of mayors in at least two major American cities (Nathe, 1983).

Natural vs. technological hazards: a five-fold distinction
One of the most detailed recent efforts to distinguish between natural and technological hazards has been made by Baum *et al.* (1983*a*). This scheme, which attempts to explain why a technological accident which produced little physical

damage may have a long, dramatic effect on public thinking, makes a distinction between natural and technological hazards based on five variables. Natural disasters are judged to create property damage that is highly visible and are predictable to the extent that they occur on a recurring basis or that accurate forecasting is possible. They are said to have a low point, when the 'worst is over', and after which the environment is restored. The effects of natural disasters are believed to extend only to the immediate victims. Based on available evidence it is concluded that natural disasters do not have extensive negative psychosocial effects beyond the immediate period of impact and recovery (e.g. Gleser *et al.*, 1981).

Technological hazards are judged by Baum *et al.* (1983*a*) sometimes to cause visible damage but also sometimes to cause invisible damage (e.g. radiation or toxic wastes). They are, as a group, believed to be less predictable than natural disasters because 'technology is not built to break down' (p. 341). While some technological catastrophes are believed to have a low point (e.g. train and marine spill accidents), some of the most stressful are believed not to have one. Incidents in which people believe that they have been exposed to toxic substances such as in the case of Love Canal, New York, and Three Mile Island, Pennsylvania, are examples of such events. The possibility of long-term health effects and delays in clean-up efforts are thought to produce strong feelings of victimization and a judgment of continuing threat (Baum *et al.*, 1981; Baum *et al.*, 1983*b*).

Finally, it is held that technological catastrophes have effects that extend beyond the immediate victims. The loss of confidence and credibility engendered by a technological catastrophe may affect people who are not directly victimized. As Slovic *et al.* (in press) contend, the societal impact of an accident is largely determined by conclusions about what it signifies or portends. An accident that causes little harm may nevertheless have immense consequences if it increases the judged probability and seriousness of future accidents. The societal impact of the reactor accident at Three Mile Island, for example, is held by some to far exceed the possible physical damage that may have occurred (*EPRI Journal*, 1980; Evans and Hope, 1982). Because of these characteristics the effects of technological catastrophes are expected to be chronic for many individuals.

Critique of the natural/technological distinction
A number of classification schemes categorizing hazardous environmental events on the basis of single descriptive characteristics have been proposed. These include the classification of hazards according to source (e.g. industry, auto emissions), use (e.g. medical X-ray), potential for harm (e.g. explosives), population exposed (e.g. asbestos workers), environmental pathways (e.g. air pollution) or varied consequences (e.g. cancer, property loss) (Hohenemser *et al.*, 1983*a*). The distinction between natural and technological causes of hazards is the most commonly used of these distinctions.

The Baum *et al.* (1983*a*) classification scheme shares both the advantages and disadvantages of earlier uses of the distinction between natural and technological hazards. The advantage is that it provides a simple, easily understood (at a superficial level) classification that has a considerable degree of intuitive appeal. The major disadvantages are that the simple distinction of natural and technological hazards does not match adequately the complexity of real life events. Its application forces the categorizer to ignore inconsistencies and to use biased examples.

The five dimensions discussed by Baum *et al.* reasonably reflect important variations among hazardous events. They do not, however, neatly collapse into the two-fold distinction of natural and technological hazards. Examples are often selected to fit the categorization scheme and important exceptions are ignored. While the overlapping range of characteristics of both classes of hazards is recognized, emphasis is given to those events that are most distinctively unlike the selected examples of the other class. Also unrecognized is a bias in the disasters typically studied and hence a bias in the information available about hazards.

Most U.S. research on natural disasters has investigated events that have short-term impacts, usually of several days to a few weeks duration, that were modest relative to local and national resources and that are usually well contained by respond-ing social agencies (Kreps, 1981). Not all natural disasters have these characteristics. Nor do all natural hazards have a sudden onset, a rapid period of impact and a low point at which the hazardous conditions begin to 'turn around'. Prolonged droughts resulting in desertification of large geographic areas would be one exception to these characteristics. The El Nino phenomenon that has produced large negative effects on fisheries both in South and North America is another example. A third example is large-scale extreme climatic change induced by the injection of materials into the atmosphere by volcanic activity or meteorite impact. Baum et al., reasonably enough, exclude evidence that natural disasters may have long-term negative effects because these studies typically lack base-rate data or control groups. Yet, similarly flawed data is taken as evidence for the long term effects of technological catas-trophes.

The distinction between natural and technological hazards has a long history both within the scientific literature and among relevant regulatory organizations. It is rooted in the history of the science of hazard research and reflects the historical evolution of government agencies and the various sciences. The scientific disciplines dealing with natural hazards, such as geology, geography and meteorology, are older than those such as toxicology, epidemiology and safety engineering which study technological hazards. The distinction also results from the relative importance of the two classes of hazards at different historical times (Harriss et al., 1978). Currently, in industrialized nations, natural hazards are of relatively less import than are those rooted in technology. The impact of technological hazards in terms of fatalities, expenditures and losses is much greater. This was not always so, nor is it yet true of developing countries. In these cases hazards rooted in nature and primarily related to agriculture, food supply and settlement take precedence.

The use of the natural vs. technological dichotomy continues, even though it poses many difficulties and results in a number of inconsistencies and oversights. The distinction remains useful if its application is limited to the relatively simple identi-fication of the causes of hazardous events and is not extended to apply to the more complex subsequent stages in the hazard sequence. Classifications of causes, either the simple distinction of natural versus technological hazards or more complicated classifications of the physical environment (e.g. Rogers and Nehnevajsa, 1984), are best suited to the needs of research that aims to improve management procedures for the prediction and prevention of hazards.

Classification of Hazard Characteristics

A major purpose of classifications of hazard characteristics is the mitigation of hazards. Three approaches to the classification of hazard characteristics, distinguished by method, data, and conceptualization of characteristics, have been taken. One of these, the psychometric approach, uses psychophysical scaling methods and multivariate analytic techniques to produce quantitative representations of judgments about hazards. The other two approaches classify the physical characteristics of hazards. One of these derives general abstract descriptive categories whereas the other one develops distinctions directly on the basis of specific physical properties. For this reason the former approach will be referred to as focusing on macro characteristics, while the latter approach focuses on micro characteristics.

Psychometric ratings of hazards
Much of the work within this approach focuses on judgments of acceptable risk. The question of 'How safe is safe enough?' motivated much of the early psychometric research as the acceptability of risk seemed central to technology development, risk assessment and the balance of risk and benefits. In an initial, well-known study by Fischhoff *et al.* (1978), each participant was asked to rate either the benefit or the risk of 30 activities and technologies. Ratings were also made of the voluntariness of risk, immediacy of effect, knowledge of risk by persons exposed to it, the newness of risk, the chronic versus catastrophic nature of the risk and the severity of consequences. These dimensions have been included in classifications of hazardous technologies proposed by Starr (1969), Lowrance (1976), Litai *et al.* (1981) and others.

Results of principal component factor analyses showed that two orthogonal factors sufficiently account for the intercorrelations within the sets of risk ratings. The first factor distinguishes between new, involuntary, highly technological items that have delayed consequences (for example, nuclear power, pesticides and food preservatives) and those that are familiar, voluntary activities with immediate consequences at the individual level (for example, swimming and skiing). This factor was labeled 'Technological Risk' and relates to how well a hazard is understood. The second factor distinguished between events that were certain to be fatal to large numbers of people should an accident occur (for example, nuclear power plants) and those with less severe consequences. The second factor was labeled 'Dread Risk'.

Factor analysis of a replication of this study using a longer list of 90 hazards identified three factors (Slovic *et al.*, 1980). The first two factors were very similar to 'risk understood' and 'dread' factors identified by the first study. The third factor had to do with the number of people exposed to the risk.

Vlek and Stallen (1980, 1981) completed a psychometric investigation which used a task different from that employed by Slovic and his colleagues. Rather than supplying respondents with the component characteristics of risk, this study asked that ratings be made of the similarity of the risks of pairs of hazards. The activities, each of which was described to the respondents, included risks/benefits 'in the small' such as 'drunken driving', and 'smoking in bed', and risks/benefits 'in the large' such as 'transportation of chlorine gas' and 'mass use of pesticides'. Multidimensional scaling analyses of the categorized rank orderings indicate that a two-dimensional cognitive configuration underlies judgments concerning the activities. The primary dimension was called 'size of potential accident'. Most respondents agreed that riskiness of activity increases with increases in potential accident. The second dimension was called 'degree of organized safety'. Approximately one-half of those interviewed judged riskiness as increasing with decreases in organized safety. The others indicated that they believed that riskiness increased with more organized safety.

There is some communality between the factors identified by Vlek and Stallen and those identified by Slovic *et al.* (in press); Green and Brown (1980) also using multi-dimensional scaling of similarity judgments have obtained a two-dimensional representation similar to that of the factor analytic studies. Johnson and Tversky (1983) in a direct comparison of factor analytic and similarity representations derived a factor space quite different from those of the earlier factor analysis studies. The study included natural hazards and diseases as well as the technologies and activities used in the Slovic *et al.* study. It was also found that judgments of similarity were very different when done using characteristics supplied by the experimenter than when based on direct comparison of the hazards. The researchers suggest that the two representations may relate to different modes of operation. Factor analytic representations may relate more directly to important attitudes towards hazards. Similarity representations may relate to reactions to new risks or the evaluation of new information about old risks. Another representation of these results in terms of common and unique features of the rated hazards revealed a distinct hierarchy of four clusters. These were termed hazards (e.g. electrocution, lightning); accidents (e.g. accidental falls, traffic accidents, airplane accidents), technological disasters (e.g. nuclear accident, toxic chemical spill) and diseases (e.g. heart disease, stroke).

Apparently risk judgments are very sensitive to context effects. Even subtle differences in the way in which information is presented, for instance, has been found to produce large differences in perceived risk (Tversky and Kahneman, 1981; Slovic *et al.*, 1982). Using a free-response technique to survey public judgments about hazardous industrial installations. Earle and Lindell (1984) obtained results different from those of earlier studies using a 'closed' format. For example, concern for future generations is often rated as very important in psychometric studies but was not indicated to be so in this free response study. In another study a randomly-selected local sample gave very different responses about concerns for personal safety depending on how questions were phrased (Cvetkovich, 1982). The survey was conducted in the month following the fatalities in Chicago resulting from cyanide-filled Tylenol capsules. When asked to list what things had caused personal worry or concern during the last month, less than one per cent of the sample mentioned the intentional contamination of drugs on store shelves. Later, in the telephone interview, when asked directly if they were concerned or worried that they might purchase a contaminated drug product over 80% stated that they were very concerned. Fischhoff and MacGregor (1983) found that four formally equivalent response modes yielded very different absolute estimates of the lethality of various potential causes of death. Similar orderings of lethality for the hazards were found across the estimates of death rate, number died, survival rate and number survived, however. A more extensive discussion of differences due to variations in method is presented in Lee (1983). Observed variations in judgments of risk due to tasks, item set and analytic methods indicate that care should be taken in selecting the methods most suitable for a particular research problem.

The psychometric approach has several weaknesses (Vlek, 1984; Otway and Thomas, 1982). Some of these relate to questions of external validity and are similar to those of most 'paper and pencil' studies. The results of psychometric studies are purely descriptive and do not provide information concerning risk judgment processes. Moreover the risk judgments recorded in psychometric studies are usually taken at face value. It is likely, however, that they are subject to the vagaries of the relationship between attitudes and behavior. Surveys and field studies relating risk judgments and actual behavior, such as actions to mitigate risk, are particularly needed. A few studies of this type have been completed (Gardner *et al.*, 1982), and more are planned or in progress. Analyses typically used in psychometric studies do not yield information concerning individual judgments. Most psychometric studies have analysed mean ratings of risk correlated with mean ratings of risk characteristics across technologies. Reliable relationships identified by this form of group analysis should reflect how society responds to hazards. Such a 'between-subjects' design, however, may not yield results reflecting the judgments of individuals rating a single technology (Gardner *et al.*, 1982; Renn, 1981).

Several limitations of the psychometric approach relate to the manner in which data are usually collected. (a) Psychometric tasks often require uncommon comparisons (for example, comparing the risks of pesticides and skiing). Since people do not commonly have to choose between one or the other, the judgments may not reflect a meaningful choice for the individual (Lee, 1983). (b) Psychometric studies often require that the risks of activities are judged independently of benefits and other contextual factors. Vlek and Stallen (1980) have found that benefits are more strongly related to the judged acceptability of activities. (c) The psychometric approach has given little attention to the positive aspects of risk taking; i.e. risk taking can be fun and may be sought after if the individual believes that it can be handled.

A basic question concerning the psychometric approach relates to the underlying argument that risk is a generic aspect of human activity and that people use this generic dimension to compare activities. Another equally valid argument is that the dimensions used for the appraisal of hazards are not universal but vary greatly depending upon situational, individual, social and cultural influences. Psychometric researchers have come to reject as logically unsound the concept of acceptable risk (Slovic *et al.*, 1982; Lee, 1983). Society accepts activities and technologies, not risk. Acceptance depends on both costs and benefits of the technology and of its alternatives. As already mentioned, variations in judgments of risk may result from the incomplete specification of these factors in psychometric studies. A start towards a more adequate conceptualization of hazards has been made by several researchers. Fischhoff *et al.* (1983) have shown that the importance of hazard characteristics such as degree of controllability, evoked dread, catastrophic potential, and equity of risk/benefit distribution greatly varies in the assessment of different energy technologies. Douglas and Wildavsky (1982), Thompson (1983), Torry (1979) and Sorensen and White (1980), using an anthropological perspective, have pointed to distinct cultural differences in the selection of events identified as hazards.

As von Winterfeld and Edwards (1984) point out, the psychometric approach as implemented leads to the conclusion that establishing the acceptability of risk is equivalent to resolving conflicts about the acceptability of new technologies. Controversies may, however, involve much more than differing levels of acceptable risk. The content of technological controversies may include various combinations of factual and value conflicts ranging from differences in beliefs about statistics, data, probabilities, cost–benefit tradeoffs, distributions of risk among the population and basic social values. Based on an analysis of 162 technological controversies, von Winterfeld and Edwards (1984) have developed a two-level classification system of controversies concerning technological innovation. The analysed cases do not include

TABLE 2
A taxonomy of technological controversies from von Winterfeld and Edwards (1984)

Technological category	Impact subcategory
A. Food/drug/consumer products	A.1. Dramatic, unexpected effects (e.g. thalidomides) A.2. Uncertain, low dose health effects (e.g. cyclamates)
B. Industrial development	B.1. Local, large scale development impacts (e.g. strip mining) B.2. Diffuse, widespread low dose pollution (e.g. transportation noise)
C. Technological mysteries and	C.1. Disaster threats and value threats catastrophe potential (e.g. LNG siting) C.2. Value threats and moral imposition (e.g. weather modification)

those technological controversies that ended in earlier failures (e.g. atomic aircraft) or those that survived to be fully implemented. The technological controversies classified according to three dimensions: (a) stakeholders — proponents (e.g. large scale developers, consumer product manufacturers) and opponents (e.g. environmentalists or consumer advocates); (b) values; and (c) pattern of conflict (e.g. risk/benefit trade-offs, product safety). The resulting three classes of controversies are further subdivided by types of impacts (e.g. dramatic, unexpected health effects vs. uncertain, low dose health effects). The entire classification scheme is outlined in Table 2 with examples.

Despite its weaknesses, the psychometric approach has yielded important generalizations about risk judgments (Slovic *et al.*, 1983). Among these are the finding that perceived risk can be quantified and that the psychometric method is well suited for identifying differences and similarities among groups. For example, technical experts tend to judge the risk of an activity relative to estimates of annual fatalities. When asked to do so non-experts can make estimates of annual fatalities having a rank order similar to those produced by experts. The estimates of risk made by lay people, however, are also sensitive to other factors, such as catastrophic potential, severity of the evoked hazard image and frequency of exposure to information about the hazard. This difference can result in a conflict between the risk judgments of experts and lay people. Psychometric studies also suggest the inadequacy of uni-dimensional indices of risk such as annual probability of death or reduction of life expectancy, measures often recommended by technical experts. People's judgments often appear to be based on several dimensions and include qualitative as well as quantitative characteristics.

Micro characteristics

In an effort to classify technological hazards and simplify analysis and management, the Hazard Assessment Group at Clark University and associated researchers (Harriss *et al.*, 1978; Hohenemser *et al.*, 1983a,b) have sought to identify common characteristics by analysing physical, biological and social descriptors. Within the conceptual framework used by this group, hazards are defined as threats to humans and what they value, and risks are quantitative measures of hazard consequences. Technological hazards are conceived of as the result of events leading from human needs and wants, to the possible release of materials and energy, to human exposure and finally to harmful consequences.

Ninety-three technological hazards were described on 12 quantitative scales including intentionality, spatial extent and concentration of impact, recurrence, and population at risk. Inspection of the scored characteristics of technological hazards lead the Clark group to conclude that a division between energy and materials hazards is an important distinction leading to four striking differences. (1) Energy releases are short, averaging less than a minute; materials releases persist for a week or more, on average. (2) Energy releases have immediate consequences with delays between exposure and consequence lasting less than a minute; materials releases have exposure-consequence delays averaging 1 month. (3) Energy releases have minor transgenerational effects; material releases affect an average of at least one future generation. (4) Energy hazards have little potential non-human mortality; material hazards can cause significant nonhuman mortality.

Factor analysis showed that five factors could be used to describe the hazards. The first, called 'Biocidal', loaded highly on experienced and potential non-human mortality and intentionality scales. Factor 2, 'Delay', loaded on persistence, delay and transgenerational effects scales. Factor 3, 'Catastrophic', loaded on the recurrence and human mortality scales. Factor 4, 'Global', loaded on population at risk and concentration scales. And Factor 5, a residual factor, loaded on the spatial extent scale. The factor structure did not change appreciably when hazards were added to or subtracted from the analysis. The identification of hazards having extreme scores on only one factor lead to the development of a seven-class taxonomy with three major groupings. This taxonomy is presented in Table 3.

Hohenemser et al. (1983a) state that a hazard classification can be evaluated on three criteria: does it describe the essential elements that make specific hazards threatening to humans and what they value? does it reflect the concern of society? does it offer a new tool for managing hazards? They suggest that they will wait the verdict of others as to whether the first criterion is met. (We would argue that this question can not be answered unless a specific purpose for the classification is stated.) Evidence that the classification satisfactorily describes the essential elements that influence judgments of hazardousness are provided by a small psychometric study of judged risk. Modest correlations between risk assessments and the descriptor scores were found in nine of the 12 cases. Modest correlations were also found between judged risk and the five identified factors. When mean ratings from the

TABLE 3
A seven class taxonomy (from Hohenemser et al., 1983)

Class	Examples
Multiple extreme hazards	Nuclear war, recombinant DNA, pesticides
Extreme hazards	
Intentional biocides	Chain saws, antibiotics, vaccines,
Persistent teratogens	Uranium mining, rubber manufacturing
Rare catastrophes	LNG explosions, commercial aviation
Common killers	Auto crashes, coal mining
Diffuse global threats	Fossil fuel, SST
Hazards	Saccharin, aspirins, appliances, skateboards, bicycles

sample were substituted for the descriptor scores, correlations with judged risk increased.

Three examples of how the classification could be used to improve hazard management are presented. The classification system can be used for comparing hazards and thereby aiding decisions concerning choices among competing technologies. The advantage of this classification is that it provides for multidimensional comparisons. A second example of managerial use is the early identification of new technologies that are likely to be problematic. Development of a classification profile would allow comparison with other hazards having similar profiles. This would warn managers of problems that may be otherwise unexpected, as well as provide them with precedents for dealing with the problems. Finally, the classification system could provide a means for identifying hazards worthy of special attention. Hazards with multiple extreme scores could be given extraordinary attention; distinctive efforts could be given to those with extreme scores and others could be dealt with through routine responses.

The Clark group classification, as any classification must be, is obviously limited in certain respects. It is intentionally limited to technological hazards, hazards are described only by their negative characteristics and not by their positive ones and the array of hazards does not include those posed by the accumulation of materials that are innocuous individually but deadly in sum. Critics have concluded that for these reasons the classification is limited as a guide for policy decisions (Rossin, 1983; Porter, 1983). The researchers recognize these limits (Hohenemser, 1984), but believe that they are not now resolvable. It is argued that the benefits of technology are much more difficult to measure than are costs because they inevitably lead to questions of values and conflicts as to who profits from a technology. The measure of the accumulative effects of hazardous material is also difficult. The Clark group has concluded that possible measurement solutions, such as adding up the estimated consequences of individual materials exposure or estimating the cumulative impact through global measures, such as employment in industry, are either presently impossible to accomplish or have large ranges of error (Harriss et al., 1978).

The Clark group effort comes closest in its approach to the procedures commonly employed by numerical taxonomy (Sokal, 1974; Snead and Sokal, 1973). Numerical taxonomy, initially developed as a general approach to biological classification, has had substantial subsequent influence in other physical and social sciences (Sokal, 1974). A basic aspect of numerical taxonomy is its use of multidimensional analysis on large arrays of characteristics. This approach as applied by the Clark group might be considered by some to be the most 'objective' approach to classification since it is based on the physical characteristics of hazards. It should be noted that the values of the hazard characteristics consist of ratings. Also the Clark analysis treats all events within classes as being represented by a single value. This assumption is necessitated by the need to reduce observations. In essence the events are presorted and the derived classification is based on an abstracted categorization of hazards. An ideal approach would be an automated procedure for collecting information about single events on a large number of variables—a technological innovation that does not seem possible within the foreseeable future.

Macro characteristics
Perry and his associates (Perry, 1982, 1983a; Perry et al., 1980) have proposed a

categorization of the physical characteristics of events for the purpose of improving the management of programs seeking compliance to warnings and evacuation programs. The classification is developed on the basis of an analysis of responses to frequently occurring natural events such as floods, tornadoes, hurricanes and earthquakes. It is suggested that the classification scheme, which is grounded on an extensive body of sociological literature on collective disaster responses (e.g. Drabek, 1969; Quarantelli, 1980a,b), can be used to generalize the findings for well-studied events to those that have either occurred infrequently or have never occurred before. The two particular events of concern are nuclear power plant accidents (Perry, 1983a) and a nuclear attack (Perry, 1982; Perry et al., 1980). It is argued that assessment of the likelihood of compliance to warnings and evacuation orders can be made by applying the classification system to make a careful analytic comparison between these events and natural hazards on the basis of the events' macro characteristics.

The logical basis of this classification scheme begins with an analysis of the definition of a disaster. It is concluded that people think of both catastrophic events associated with the forces of nature such as floods, earthquakes or hurricanes and other events such as explosions, industrial accidents, wars or nuclear attack as appropriately labeled with the umbrella term 'disaster'. Similarly the scientific use of the term is inclusive of all of these events. It is generally accepted that a disaster is an event that occurs at a particular time and place, that places a large group of people in severe danger, that disrupts at least some of a society's essential functions and in which large numbers of individuals incur losses (Fritz, 1961; Barton, 1970).

Following earlier work by Anderson (1969) and Barton (1970), Perry specifies similarities and differences among classes of events with regard to general disaster characteristics. Five characteristics are selected for use in the comparison. (1) Scope of impact is a geographic reference to the extensiveness of the event. An event with a narrow impact encompasses a small area and a small number of people. An event with a widespread impact effects a wide area and a large number of people. (2) Speed of onset indicates the rapidity with which the event occurs or the amount of time between detection of a hazard and its effect on a community. (3) Duration of impact is the time that passes between initial impact and when the effects of the hazard subside. (4) Social preparedness is the extent to which the current state of technology allows the anticipation or prediction of a threat. (5) Secondary impact refers to the consequences produced by the hazard in addition to those occurring at onset. Examples of secondary impacts are radiation damage from a reactor accident, aftershocks following an earthquake and silt deposits from a river flood. A comparison on the five dimensions of the hazards of riverine floods, volcanoes, earthquakes, hurricanes, nuclear power plant accidents and nuclear attack is given in Table 4.

One major shortcoming of the macro-classification is that human responses are not a direct function of an event's physical characteristics, rather they are more directly related to *judged* characteristics and contextual factors. Platt (1983) raises the question of whether nuclear war, because of its potential apocalyptic impact and other unique characteristics, is comparable to a natural disaster or is a wholly different phenomenon. While Perry (1983a,b) argues that a variety of disasters can be profitably examined within the same definitional and conceptual scheme, he recognizes that unique hazard characteristics may have important consequences for human responses. For example, public opinion surveys indicate that the judgments

TABLE 4

Hazard agents classified by defining characteristics of disaster (from Perry, 1983 and Perry et al., 1980)

Defining characteristics	Hazard agent					
	Riverine flood	Earth-quake	Hurri-cane	Vol-canoes	Nuclear power plant accident	Nuclear attack
Scope of impact	Narrow	Widespread or narrow	Widespread	Variable	Variable	Widespread or narrow
Speed of onset onset	Sudden or gradual	Sudden	Gradual	Sudden	Sudden	Sudden
Duration of impact	Short (but variable)	Short or repeated	Short (but variable)	Long	Long	Short or repeated
Secondary impacts (damage or public health)	Public health problems; some physical damage	Both	Both	Both	Public health (radiation)	Both (radiation)
Social pre-paredness (predict-ability)	Yes	Not at present	Yes	Detect but not predicted	Detect but not predicted	Yes, but limited

of many Americans concerning nuclear reactors will be quite different from those concerning other hazards (Lindell *et al.*, 1978). The nuclear issue is one that engenders a very strong emotional response in many. The public views nuclear power as being especially hazardous, having the potential to produce very extensive long term irreversible primary and secondary effects. Recognition of these or other differences does not preclude comparisons using the proposed classification scheme. It only indicates the need to recognize the operation of non-categorized dimensions and to remember that qualifications must be made when attempting to generalize from responses to infrequent or 'never-occurring-before' events.

Hazard managers are charged with the duty of taking actions to reduce the risks involved in hazardous situations. These actions should be based on the best available information. For certain hazards the only available information may be that which can be generalized from other events on the basis of a classification scheme. It is assumed that the development of a comprehensive classification of hazardous events would permit the development of management techniques that could be used in a variety of disasters (Perry, 1983a). These techniques, termed generic functions, presumably represent general tasks that have to be accomplished in a variety of disaster situations, without regard to the nature of the agent causing the event. Examples of generic functions include emergency medical care to victims, finding and rescuing people, sheltering those in need, and reuniting separated family members. As Perry's classification demonstrates, it is possible to develop a useful general categorization across various specific disasters which includes such responses.

Classification of Responses

Individual responses

It is generally recognized that being a victim of a hazardous event may result in severe psychological reactions. These reactions, collectively termed 'post-traumatic stress disorder' are described by the American Psychiatric Association (1980) in its *Diagnostic and Statistical Manual of Mental Disorders* as being of three types. Reactions to a recognizable stressor event include: (a) re-experiencing the trauma through memories, intrusive thoughts, or dreams; (b) a numbing of emotional responsiveness which is expressed through a feeling of detachment from others, constricted affect or diminished interest in previously significant activities; and (c) other symptoms including an exaggerated startle response, sleep disturbances, guilt, memory impairment or trouble concentrating, and phobias about activities triggering recollections about the event. Recent years have seen a significant increase in systematic research and conceptual efforts to answer the obviously significant basic questions about stress and hazardous events such as the frequency and duration of these reactions (e.g. Quarantelli, 1980*b*; Lystad, 1984; Ahearn and Cohen, 1984). One of these efforts has been the development of a taxonomy of environmental stressors by Lazarus and Cohen (1977).

According to Lazarus and colleagues (Lazarus, 1966; Lazarus and Launier, 1978; Coyne and Lazarus, 1982; Folkman and Lazarus, 1982) stress should be viewed as a special transaction or relationship between two systems. The systems may be within the same individual (e.g. id, ego, superego), two or more different individuals, or an individual and the environment. The third case is obviously the one of concern here. A stressful environmental transaction is one which demands, taxes or exceeds the individual's resources. Three conventional measures are used to assess the individual's reactions: (a) somatic or physical reactions to stress; (b) behavioral reactions, including an individual's coping behavior, disorganized functioning and expressive behavior; and (c) subjective reactions which are the individual's feeling of experiencing stress.

This model emphasizes that the person is in a continual process of appraising the significance of the environment for personal well being. Primary appraisal seeks to answer the question 'Am I okay or in trouble?' If the appraisal results in the conclusion that an event possesses the possibility of harm, threat or challenge, conditions that tax the individual's resources, psychological stress occurs. Secondary appraisal addresses questions concerning what can be done about a stressful encounter. The appraisal processes lead to coping actions that are consistent with the individual's needs and goals. The effects of coping are in turn appraised and reacted to as part of the continuous interaction between the individual and the environment. Research on categories of stressors focuses on adjustments generated in response to environmental conditions, the control of stressors, the positive and negative value attributed to environmental stressors, and the adjustment patterns within each individual. Using this conceptualization of stress, Lazarus and Cohen (1977) develop a classification that generalizes environmental stress conditions into three groups according to their speed of onset (sudden vs. gradual change), whether the experience is shared or individual and whether the experience with the stressor is of short or long duration.

The first group in the classification are cataclysmic events which affect large numbers of people and are sudden, unique, powerful single life events over which the individual has little or no control. Cataclysms include natural occurrences such as tornadoes and flash floods (Green and Gleser, 1983) as well as human made events such as the bombing of cities and mass relocation of populations (Trimble,

1980). Although not explicitly mentioned, conceivably cataclysms would also include threats generated by the concerns about technological hazards such as EDB contamination at Times Beach, Missouri, and the nuclear plant accident at Three Mile Island. Cataclysms would definitely include what Lifton (1967) has called historical crises, such as the atomic bombing of Hiroshima and Nagasaki, events which have a large psychological impact on the individuals experiencing them and which mark a change or turning point for society at large. As these examples show, the categorization assumes that there is not a unique psychological stress response to technological and natural hazards. Lazarus and Cohen do recognize, however, that coping behaviors may be very specific to particular events. Behaviors for coping with a flood are quite different from those for dealing with bombing.

The second category of hazards are life events which are powerful enough to challenge adaptive responses in the same way that cataclysms do, but which affect fewer people at a single time. This category includes death of a loved one, divorce, illness, loss of one's job, and being robbed, assaulted or raped. Much recent social psychological research examining adaptation to stressors has focused on events within this class (e.g. Janoff-Bulman and Frieze, 1983). This work suggests that becoming a victim to one of these events forces the individual to re-examine very basic assumptions about life. Among these are the assumptions of invulnerability to negative experiences and that life is meaningful and the generally positive self concept that most individuals have.

The third category, 'daily hassles', includes chronic, repetitive stressors of everyday life. The key elements distinguishing daily hassles from the other two categories of events is that they are stable rather than temporary or intermittent in nature. They also are mundane and, contrary to the nature of cataclysms and stressful life crises, are usually non-urgent and commonplace. Any single exposure to events in this category are not generally as stressful as cataclysms or powerful life events. Examples of daily hassles are living in poverty, ordinary family decision conflicts, sex role conflicts, and the overload or underload of assembly line work. Examples of daily hassles produced by the physical environment given by Lazarus and Cohen include excessive noise or air pollution and rush-hour traffic.

Recently, Campbell (1983) has argued that a fourth category, ambient stressors, should be distinguished from daily hassles. Characteristics of ambient stressors are that they are chronic, negatively valued, non-urgent physically perceptible and intractable to the efforts of individuals to change them. Ambient stressors include global environmental conditions such as urban noise and air pollution. Campbell suggests that these global conditions are more or less continuous, contrary to the suggestion by Lazarus and Cohen that daily hassles are discrete events. Coping responses to ambient stressors may be extreme, as opposed to the commonplace efforts usually taken in dealing with daily hassles. For example, noise could be reduced by moving one's residence to a quieter location. The research evidence reviewed by Campbell concerning psychological appraisals of stressors supports the contention that the distinction between ambient stressors and daily hassles is a valid one.

Lazarus and Cohen have certain reservations about the term environmental stress since it suggests that the source of stress is external to the person without considering the important determinants the individual brings to the stress response. Instead, they suggest that stress be defined as a troubled commerce with the environment. This transactional perspective raises questions about the adequacy of conceptualizations of stress taking a solely individualistic point of view. The transactional perspective also raises important issues for classification systems that include stress and other psychological responses to hazards. These issues include the following (Lazarus and Cohen, 1977; Lazarus and Launier, 1978; Coyne and Lazarus, 1982; Folkman and Lazarus, 1984): (a) the need to define stressor events as to their type, severity, and duration; (b) consideration of the duration of the stress response of its pattern; (c) consideration of the generality of the stress response; (d) consideration of the 'temporal dimensions' of the stressor–stress responses association; and, most importantly, (e) systematic concern with processes mediating between the environment and the stress response.

There appears to be wide variability in how individuals appraise the hazardousness of events and cope with those that are judged to be threatening and stressful. Research identifying the frequency of responses and the reasons for their occurrence has just begun (e.g. Quantelli, 1980*b*; Silver and Worthman, 1980; Janoff-Bulman and Frieze, 1983). To address the issues raised by the Lazarus–Cohen classification scheme, longitudinal research focusing on individual judgment processes and behavioral adaptations is needed.

Aggregate responses

Disaster research. Kreps (1982), working out of the same sociological tradition as Perry, has developed a classification of social units and responses to disasters. This taxonomy recognizes the necessity of specifying the physical and temporal dimensions of impact. Like the Perry classification, it is based on the definition of a disaster as consisting of events that can be designated in time and space and which have impacts on social units that enact responses to the impacts (Fritz, 1961). The properties of each of the characteristics are also described in terms of physical, temporal and social dimensions. Interpretation of impacts of disasters include the four characteristics of magnitude of impact: scope of impact, length of forewarning and duration of impact. Four elements are used to define the patterns of organizational response to a disaster (Kreps, 1982; 1983).

One of these is the specification that a function or area of activity is the domain of a particular agency. Examples of this element are the official declaration that evacuation will be carried out by the police department or that the National Guard will protect homes from looting following a tornado. A second element, actions, consists of the specific behaviors needed to respond effectively to a disaster—for example, the actual movement of evacuees from one area to another or the patroling of the streets. A third element is the development of a task structure that involves the organization of the response activities. Organizing the logistics of finding, dispatching and fueling evacuation vehicles and assigning personnel to operate them are examples of this element. The fourth element consists of the human and material resources needed to respond to a disaster. Evacuation usually involves the need for trucks, boats or other vehicles, drivers and dispatchers. The taxonomy is defined by the temporal order of occurrence of the elements. It is generated by the basic factorial design of all possible combinations of all four elements in each temporal order and the forms involving one, two and three of the elements.

Kreps (1982) has examined archival data from 423 instances of organization occurring during 15 selected natural disasters (e.g. earthquakes, floods, hurricanes, tornadoes). The studies yielding the data were collected by the Disaster Research

Center of the Ohio State University. The data collection was slanted towards collecting information about organizations existing before a disaster. Even so, a pattern in which a designated organization uses its resources according to a developed plan accounts for only 39% of the total patterns identified. Fifty-two of the total 423 instances were enacted by social units that did not exist prior to the disaster. Kreps (1982) concludes that disasters induce rapid and extensive departures from established social routines and that one can not assume that pre-disaster domain-initiated patterns are always the most appropriate or effective.

Hazard research. Mileti (1980) proposes a classification that covers the theoretical range of aggregate responses which reduce the risks of extremes in the physical environment. The classification is based on distinctions made by human ecologists in the mechanisms by which humans adapt to routine environments. Four mechanisms have been identified (Micklin, 1973): (1) engineering mechanisms which include technological inventions; (2) symbolic mechanisms which include culture; (3) regulatory mechanisms that define public policy and social control; and (4) distributional mechanisms that specify the movement of people, activities and resources. The Mileti classification cuts across these mechanisms and differentiates three primary classes of behavior: purposeful adjustments, incidental adjustments and unwitting adaptations. Each of these is further divided into two to three subclasses. Purposeful adjustments consist of actions that involve choice and change such as choosing to move residential location following a disaster, behavior that reduces losses such as the development of prediction and warning systems, and behaviors resulting in the redistribution of loss such as insurance, disaster relief and charity. Incidental adjustments include choices, reduction of loss and redistribution of loss that are not directly related to hazards. For example, choices might be influenced by non-hazard-specific land codes, losses could be reduced by fire codes and improvements in fire fighting techniques, and saving accounts could be a source of resources for redistributing disaster losses. Unwitting adaptations consist of (as yet unidentified) biological changes and cultural adaptations such as redistribution of populations (e.g. de-urbanization), shifts in family structure and reductions in wealth.

The Mileti classification serves the useful purpose of providing an orderly extension of the analysis of adaptive mechanisms from routine environments to the boundaries of environmental extremes. Furthermore, the classification identifies important gaps in knowledge. Little is known about the interactions among adjustments. It is usually assumed that the adoption of one adaptive mechanism affects the adoption of others. There is little empirical evidence to support this assumption. The research review necessitated by development of the classification indicates that a large array of different independent, dependent and mediating variables related to the adoption of aggregate adaptive behaviors has been studied. No attempts have been made to determine the relative importance of alternative proposed explanations within a unified theoretical model. Also, limited existing theory focuses almost exclusively on the adoption of risk-mitigating policy, not on policy implementation. If a risk-mitigating policy is adopted but is not implemented, risk remains unaffected. Thus, scientific understanding of the effectiveness of adaptive mechanisms is greatly restricted if implementation is ignored. Adoption of risk-mitigating policies should be conceived of as a mediating variable between judged risk and the implementation of policies.

Conclusion

Classification is a method of information management that simplifies and orders the complex and diverse information available about hazards, promotes effective communication, and improves analytic thinking. The recent proliferation of hazard classifications suggests that these advantages are readily apparent to researchers and others. The information management perspective developed in this paper suggests that there is now a need to integrate the diverse classifications into a single general framework. There are four advantages to using the life history of hazard framework for this integrative function.

First, the life history organization not only brings order to what was otherwise a confusion of different approaches, but it also identifies gaps in knowledge about hazards. Very little is known about the interaction of the variables in the hazard process model. Most researchers have been ready to assume that physical characteristics influence risk judgments, that hazard images relate to hazard responses and that individual images and responses are related to aggregate responses. Empirical research specifying how these interactions take place, how they change over time and how they may be influenced by contextual factors is lacking. Answering questions concerning the dynamic relationships of the hazard process demands a new agenda of substantive questions and the adoption of a variety of research methods not now extensively used. Approaches suggested by this review include studies that make cross-hazard comparisons, longitudinal panel studies investigating adaptation to hazards across time, general population surveys appropriate for providing baseline information for the study of the effects of victimization by hazards, and laboratory studies that synthesize and control the complexities of real world conditions. Should a research agenda emphasizing the dynamics of the hazard process be adopted, the nature of classification will be changed. A dynamic perspective encourages the integration of classifications at different steps in the hazard model and forces attention away from static variables.

Second, the natural history of hazard framework represents a multidimensional classification structure. As shown by attempts to distinguish natural from technological hazards, classifications which attempt to organize events on the basis of a single defining characteristic which is uniform among members of each class are usually inadequate to reflect the descriptive complexity required by classification users. Such monothetic classifications are too simple to reflect the diversity of details researchers usually wish to recognize. Their application may demand that users be inconsistent and/or selective in the use of examples and other evidence. Polythetic classifications in which classes are distinguished on the basis of a large number of characteristics and in which members of a class share a large proportion of their characteristics but not necessarily all of them are preferred (Beckner, 1959; Ainsworth and Sneath, 1962).

Third, the natural history of a hazard framework represents hazards as transactions between humans and the environment. It follows that classifications of interest to environmental psychology cannot be solely based on either the physical properties of events or subjective impressions of event characteristics. Classification systems should assume a transactional perspective and recognize the relationship between physical properties and psychosocial processes. The life history approach indicates that concern over physical vs. subjective representations of the environment is a false issue. Either side in the debate is an incomplete view of the total human–environment interaction.

Fourth, the life history of hazard framework provides a structure for organizing practical information as well as scientific information about hazards. In a discussion of the needed agenda for risk decision making, Lowrance (1983; p. 6) states that 'The central methodological task is to develop ways of intercomparing risks against risks, weighing risks against benefits they accompany, and appraising the societal return from reducing risks.' As this review shows, a number of classifications have been motivated by such practical purposes and all of the systems have some applied relevance. Reviews of classifications aimed at improving decision making of risk managers provide very little guidance concerning application to practical situations (see for example, Solomon *et al.*, 1982). The utility of the life history organization is that it arrays the classifications at each step in the hazard process so that general groups of practical responses to hazards become apparent. The first step, focusing on hazard causes in the outer physical world, identifies practical efforts for directly identifying, reducing or avoiding negative effects by prevention and prediction. The second step, focusing on the characteristics of hazards, identifies a group of practical efforts including education and communication of information about hazards. The third step, focusing on responses to hazards, identifies practical efforts such as programs to deal with the negative psychological consequences of hazards and the adoption and implementation of plans to respond to disasters and to avoid future hazards. Thus, the life history organization aids the selection of classifications appropriate to specific risk management tasks.

Acknowledgement

Preparation of this article was supported in part by National Science Foundation Grant PRA-8312309.

References

Ahearn, F. and Cohen, R. (1984). *Annotated Bibliography on Mental Health Research in Disasters*. Washington, D.C.: U.S. Government Printing Office.
Ainsworth, G. C. and Sneath, P. H. A. (1962). *Microbial Classification*. Cambridge: Cambridge University Press.
American Psychiatric Association (1980). *Diagnostic and Statistical Manual of Mental Disorders* (DSM III), 3rd Edit. Washington D.C.: American Psychiatric Association.
Anderson, W. A. (1969). Disaster warning and communication in two communities. *Journal of Communication*, **19**, 92–104.
Barker, R. G. (1968). *Ecological Psychology: Concepts and Methods for Studying the Environment of Human Behavior*. Stanford, California: Stanford University Press.
Barton, A. (1970). *Communities in Disaster*. New York: Anchor Books.
Baum, A., Singer, J. E. and Baum, C. S. (1981). Stress and the environment. *Journal of Social Issues*, **33**, 4–35.
Baum, A., Fleming, R. and Davidson, L. M. (1983a). Natural disaster and technological catastrophe. *Environment and Behavior*, **15**, 333–354.
Baum, A., Fleming, R. and Singer, J. E. (1983b). Coping with victimization by technological disaster. *Journal of Social Issues*, **39**, 117–138.
Beckner, G. G. (1959). *The Biological Way of Thought*. New York: Columbia University Press.
Berlin, B., Breedlove, D. E. and Raven, P. H. (1969). Covert categories and folk taxonomies. *American Anthropologist*, **70**, 290–299.

Bulman, R. J. and Wortman, C. B. (1977). Attributions of blame and coping in the 'real world': Severe accident victims react to their lot. *Journal of Personality and Social Psychology*, **35**, 352–363.

Burton, I., Kates, R. W. and White, G. F. (1968). *The Human Ecology of Extreme Geographical Events* (Natural History working paper No. 2). Toronto, Ontario: Department of Geography, University of Toronto.

Campbell, J. M. (1983). Ambient stressors. *Environment and Behavior*, **15**, 355–380.

Committee on Socioeconomic Effects of Earthquake Predictions (1978). *A Program of Studies on the Socioeconomic Effects of Earthquake Predictions.* Washington, D.C.: National Academy of Sciences, National Research Council.

Coyne, J. C. and Lazarus, R. S. (1982). Cognitive style, stress perception and coping. In I. L. Kutash and L. B. Schlesinger (eds), *Handbook on Stress and Anxiety.* San Francisco: Jossey-Bass.

Craik, K. (1970). Environmental psychology. In K. Craik, B. Kleinmuntz, R. Rosnow, R. Rosenthal, J. A. Cheney and R. H. Walters (eds), *New Directions in Psychology*, Vol. 4. New York: Holt, Rinehart and Winston.

Cvetkovich, G. (1982). Comparison of free and closed-ended response modes for assessing public fear. Unpublished raw data. Western Washington University.

Drabek, T. (1969). Social processes in disaster. *Social Problems*, **16**, 336–347.

Dunlap, R. E. and Catton, W. R., Jr. (1983). What environmental sociologists have in common (whether concerned with 'built' or 'natural' environments). *Sociological Inquiry*, **53**, 113–135.

Douglas, M. and Wildavsky, A. (1982). How can we know the risks we face? Why risk selection is a social process. *Risk Analysis*, **2**, 49–52.

Earle, T. C. and Cvetkovich, G. (1983). *Risk Judgment and the Communication of Hazard Information: Toward a New Look in the Study of Risk Perception.* Seattle, Washington: Battelle Human Affairs Research Centers.

Earle, T. C. and Lindell, M. K. (1984). Public perception of industrial risks: a free-response approach. In R. A. Waller and V. T. Corello (eds), *Low Probability/High Consequence Risk Analysis: Issues, Methods and Case Studies.* New York: Plenum.

EPRI Journal. Assessment: The impact and influence of TMI. Palo Alto, California: Electric Power Research Institute, Vol. 5, 24–33.

Evans, N. and Hope, C. W. (1982). *Costs of nuclear accidents: implication for reactor choice.* Energy Research Group Report 82/17. Cavendish Laboratory, Cambridge University.

Fiksel, J. (1981). Stress and stability: New concepts for risk management. *Human Systems Management*, **2**, 26–33.

Fisher, J. D., Bell, P. A. and Baum, A. (1984). *Environmental Psychology*, 2nd Edit. New York: Holt, Rinehart and Winston.

Fischhoff, B. and MacGregor, D. (1983). Judged lethality: How much people seem to know depends upon how they are asked. *Risk Analysis*, **3**, 229–236.

Fischhoff, B., Slovic, P., Lichtenstein, S., Read, S. and Coombs, B. (1978). 'How safe is safe enough?' A psychometric study of attitudes towards technological benefits. *Policy Science*, **8**, 127–152.

Fischhoff, B., Watson, S. R. and Hope, C. (1983). Defining risk. Unpublished manuscript. Eugene, Oregon: Decision Research.

Folkman, S. and Lazarus, R. S. (1982). Stress and coping theory applied to the investigation of mass industrial psychogenic illness. In M. J. Colligan, J. W. Pennebaker and L. R. Murphy (eds), *Mass Psychogenic Illness: A Social Psychological Analysis.* Hillsdale, New Jersey: Erlbaum.

Freedman, J. L. and Sears, D. O. (1965). Selective exposure. In L. Berkowitz (eds), *Advances in Experimental Social Psychology*, Vol. 2, pp. 57–97. New York: Academic Press.

Frey, D. and Wickland, R. A. (1978). A clarification of selective exposure. *Journal of Experimental Social Psychology*, **14**, 132–139.

Fritz, C. E. (1961). Disasters. In R. Merton and R. Nisbet (eds), *Contemporary Social Problems.* New York: Harcourt, Brace and World.

Gardner, G. T., Tiemann, A. R., Gould, L. C., DeLucca, D. R., Dobb, L. W. and

Stolwijk, J. A. J. (1982). Risk and benefit perceptions, acceptability judgments and self-reported actions toward nuclear power. *Journal of Social Psychology*, **116**, 179–197.

Glass, D. C. and Singer, J. E. (1972). *Urban Stress: Experiments on Noise and Social Stressors*. New York: Academic Press.

Gleser, G. C., Green, B. L., Winget, C. W. (1981). *Prolonged psychological effects of disaster: A study of Buffalo Creek*. New York: Academic Press.

Gould, S. J. (1974). The first decade of numerical taxonomy (Review of *Numerical Taxonomy: The Principles of Practice of Numerical Classification*) *Science*, **183**, 739–740.

Green, C. H. and Brown, R. A. (1980). *Through a glass darkly: Perceiving perceived risks to health and safety*. School of Architecture, Duncan of Jordanstone College of Art, University of Dundee, Scotland.

Green, B. L. and Gleser, G. C. (1983). Stress and long-term psychopathology in survivors of the Buffalo Creek disaster. In D. F. Ricks and B. S. Dohrenwend, *Origins of Psychopathology*. New York: Cambridge University Press.

Greenwood, D. R., Kingsbury, G. L. and Cleland, J. G. (1984). *A Handbook of Key Federal Regulations and Criteria for Multimedia Environmental Control* (EPA-600/7-79-175). Washington D.C.: Environmental Protection Agency.

Hammond, K. R., Anderson, B. F., Sutherland, J. and Marvin, B. (1984). Improving scientists' judgments of risk. *Risk Analysis*, **4**, 69–78.

Harriss, R. C., Hohenemser, C. and Kates, R. W. (1978). Our hazardous environment. *Environment*, **20**, 38–40.

Hewitt, K. and Burton, I. (1971). *The Hazardousness of a Place: A Regional Ecology of Damaging Events*. Toronto: University of Press.

Hirschman, E. C. (1981). Social and cognitive influences on information exposure: A path analysis. *Journal of Communication*, **31**, 76–87.

Hohenemser, C. (1984). Letter. *Science*, **221**, 1244.

Hohenemser, C., Kasperson, R. E. and Kates, R. W. (1982). Casual structure: A framework for policy formulation. In C. Hohenmser and J. X. Kasperson (eds), *Risk in the Technological Society*. Boulder, Colorado. Westview Press.

Hohenemser, C., Kates, R. W. and Slovic, P. (1983a). The nature of technological hazard. *Science*, **220**, 378–384.

Hohenemser, C., Kates, R. W. and Slovic, P. (1983b). *A Taxonomy of Technological Hazards*. Worcester, Mass.: Center for Technology, Environment, and Development, Clark University.

Janis, I. L. and Mann, L. (1977). *Decision Making: A Psychological Analysis of Conflict, Choice and Commitment*. New York: Free Press.

Janoff-Bulman, R. and Frieze, I. H. (1983). A theoretical perspective to understanding reactions to victimization. *Journal of Social Issues*, **39**, 1–19.

Johnson, E. J. and Tversky, A. (1983). *Representations of perceptions of risks*. Stanford, California: Department of Psychology, Stanford University.

Kaplan, S. (1983). A model of person-environment compatibility. *Environment and Behavior*, **15**, 311–332.

Knott, J. and Wildavsky, A. (1979). If dissemination is the solution, what is the problem? *Knowledge: Creation, Diffusion, Utilization*, **1**, 537–1242.

Kreps, G. A. (1981). The worth of the NAS-NRC and DRC studies of individual and social responses to disasters. In J. D. Wright and P. H. Rossi (eds), *Social Science and Natural Hazards*. Cambridge, Massachusetts: Abt Books.

Kreps, G. A. (1982). *Disaster and the Social Order: Definition and Taxonomy*. Williamsburg, Virginia: College of William and Mary, Department of Sociology.

Kreps, G. A. (1983). The organization of disaster response: Core concepts and processes. *International Journal of Mass Emergencies and Disasters*, **1**, 37–48.

Lacy, H. M. (1979). Control, perceived control, and the methodological role of cognitive constructs. In L. C. Perlmutter and R. A. Monty (eds), *Choice and Perceived Control*. Hillsdale, New Jersey: Erlbaum.

Lawless, E. W. (1977). *Technology and Social Shock*. New Brunswick, New Jersey: Rutgers University Press.

Lazarus, R. S. (1966). *Psychological Stress and the Coping Process.* New York: McGraw-Hill.

Lazarus, R. S. and Cohen, J. D. (1977). Environmental stress. In I. Altman and J. F. Wohlwill (eds), *Human Behavior and the Environment: Current Theory and Research,* Vol. 2. New York: Plenum.

Lazarus, R. S. and Launier, R. (1978). Stress-related transactions between person and environment. In L. A. Pervin and M. Lewis (eds), *Perspectives in International Psychology.* New York: Plenum.

Lee, T. R. (1983). The perception of risks. In *Risk Assessment: A Study Group Report of the Royal Society.* London: The Royal Society.

Lifton, R. J. (1967). *Death in Life: Survivors of Hiroshima.* New York: Simon and Schuster.

Lind, N. C. (1984). Editorial. *Risk Abstracts,* **1**, 1–2.

Lindell, M., Earle, T., Herbert, D., and Perry, R. W. (1978). *Radioactive Wastes: Public Attitudes.* Seattle, Washington: Battelle Human Affairs Research Centers.

Linnerooth, J. Uncertainty in the policy process. *Proceedings of the Ninth Research Conference on Subjective Probability, Utility, and Decision Making* (pp. 87–106). Groningen, Netherlands.

Litai, D., Lanning, D. D. and Rasmussen, N. C. (1981). *The Public Perception of Risk.* Cambridge, Massachusetts: Department of Nuclear Engineering, Massachusetts Institute of Technology.

Lowrance, W. W. (1976). *Of Acceptable Risk: Science and the Determination of Safety.* Los Altos, California: Kaufman.

Lowrance, W. W. (1983). The agenda for risk decision-making. *Environment,* **25**, 4–8.

Lystad, M. (1984). Mental health studies of emergencies: where we are and where we need to go. *Natural Hazards Observer,* **8**, 1–2.

Meyer, M. W. and Solomon, K. A. (1984). Risk management in local communities. *Policy Sciences,* **16**, 245–265.

Micklin, M. (1973). *Population, Environment and Social Organization.* Hinsdale, Illinois: Dryden Press.

Mileti, D. S. (1980). Human adjustment to the risk of environmental extremes. *Sociology and Social Research,* **64**, 327–347.

Mileti, D. S., Hutton, J. R. and Sorenson, J. H. (1981). *Earthquake Prediction Response and Options for Public Policy.* Boulder, Colorado: Institute of Behavioral Science.

Miller, J. D. and Barrington, T. M. (1981). The acquisition and retention of scientific information. *Journal of Communication,* **31**, 178–189.

Miller, P. Y. and Fowlkes, M. R. (1984). In defense of 'man-made' disaster. *Natural Hazards Observer,* **8**, 3, 11.

Miller, J. D., Suchner, R. W. and Voelker, A. M. (1980). *Citizenship in an Age of Science.* New York: Pergamon Press.

Moos, R. A. (1976). *The Human Context: Environmental Determinants of Behavior.* New York: Wiley.

Moos, R. H. and Gerst, M. S. (1974). *University Residence Environmental Scale.* Palo Alto, California: Consulting Psychologists Press.

Nathe, S. K. (1983). Nobody was dreaming of this white christmas. *Natural Hazards Observer,* **7**, 4.

National Research Council (1982). *Risk and Decision Making: Perspectives and Research.* Washington, D.C.: National Academy Press.

Nelkin, D. (1977). *Technological Decisions and Democracy: European Experiments in Public Participation.* Beverly Hills, California: Sage Publications.

Otway, H. and Thomas, K. (1982). Reflections on risk perception and policy. *Risk Analysis,* **2**, 69–82.

Perry, R. W. (1982). *The Social Psychology of Civil Defense.* Lexington, Massachusetts: Lexington Books.

Perry, R. W. (1983*a*). *Comprehensive Emergency Management Evacuating Threatened Populations.* Seattle, Washington: Battelle Human Affairs Research Centers.

Perry, R. W. (1983*b*). Comparing disaster agents and evaluating CRP. *Natural Hazards Observer,* **7**, 6, 4.

Perry, R. W., Lindell, M. K. and Greene, M. R. (1980). *The Implications of Natural Hazard Evacuation Warning Studies for Crisis Relocation Planning.* Seattle, Washington: Battelle Human Affairs Research Centers.

Peterson, C. and Seligman, M. E. P. (1983). Learned helplessness and victimization. *Journal of Social Issues*, **39**, 103–116.

Platt, R. H. (1983). Is nuclear war comparable to a natural disaster? *Natural Hazards Observer*, **7**, 4.

Porter, A. L. (1984). Letter. *Science*, **221**, 1244.

Quarantelli, E. L. (1980*a*). *Evacuation behavior and problems.* Columbus, Ohio: Ohio State University Disaster Research Center.

Quarantelli, E. L. (1980*b*). *The Consequences of Disasters for Mental Health: Conflicting Views.* Columbus, Ohio: Ohio State University, Disaster Research Center.

Raven, P. H., Berlin, B. and Breedlove, D. E. (1971). The origins of taxonomy. *Science*, **174**, 1210–1213.

Renn, O. (1981). *Man, Technology, and Risk: A Study on Intuitive Risk Assessment and Attitudes Towards Nuclear Power* (Report Jul-Spez 115). Julich, Federal Republic of Germany: Nuclear Research Center.

Riegel, K. F. (1975). *The Development of Dialectic Operations.* Basel: Karger.

Riegel, K. F. (1976). The dialectics of human development. *American Psychologist*, **31**, 689–700.

Rogers, G. O. and Nehnevajsa, J. (1984). *Behavior and Attitudes Under Crisis Conditions: Selected Issues and Findings.* Pittsburg, Pennsylvania: University Center for Social and Urban Research, University of Pittsburg.

Rossin, A. D. (1984). Letter. *Science*, **221**, 1244.

Rothbaum, F., Weisz, J. R. and Snyder, S. S. (1982). Changing the world and changing the self: A two-process theory of perceived control. *Journal of Personality and Social Psychology*, **42**, 5–37.

Rowe, W. D. (1977). *An Anatomy of Risk.* New York: Wiley.

Silver, R. L. and Worthman, C. B. (1980). Coping with undesirable life events. In J. Garber and M. E. P. Seligman (eds), *Human Helplessness: Theory and Applications.* New York: Academic Press.

Sims, J. H. and Baumann, D. D. (1974). The tornado threat: Coping styles of the north and the south. In J. H. Sims and D. D. Baumann (eds), *Human Behavior and the Environment: Interactions Between Man and his Physical World.* Chicago, Illinois Maaroufa Press, Inc.

Slovic, P., Fischhoff, B. and Lichtenstein (1980). In R. C. Schwing and W. A. Albers (eds), *Societal Risk Assessment: How Safe is Safe Enough?* New York: Plenum, pp. 181–216.

Slovic, P., Fischhoff, B. and Lichtenstein, S. (1982). Response mode, framing, and information-processing effects in risk management. In R. Hogarth (ed.), *New Directions for Methodology of Social and Behavioral Science: Question Framing and Response Consistency.* San Francisco, California: Jossey-Bass.

Slovic, P., Fischhoff, B. and Lichtenstein, S. (1983). Behavioral decision theory perspectives on risk and safety. In *Proceedings of the Ninth Research Conference on Subjective Probability, Utility, and Decision Making* (pp. 87–106). Groningen, Netherlands.

Slovic, P., Lichtenstein, S. and Fischhoff, B. (in press). Modeling the societal impact of fatal accidents. *Management Science.*

Sneath, P. H. A. and Sokal, R. R. (1973). *Numerical Taxonomy.* San Francisco: Freedman.

Sokal, R. J. (1974). Classification: Purposes, principles, progress, prospects. *Science*, **185**, 4157, 1115–1123.

Solomon, K. A., Meyer, M. A., Szabo, J. and Nelson, P. (1982). *Classification of Risks*, (UCLA-ENG-8245). Los Angeles, California: University of California, School of Engineering and Applied Science.

Sorensen, J. H. and White, G. F. (1980). Natural hazards: A cross-cultural perspective. In I. Altman, A. Rapoport and J. F. Wohlwill (eds), *Human Behavior and Environment*, Vol. 4. New York: Plenum Press.

Starr, C. (1969). Social benefits vs. technological risk. *Science*, **165**, 1232–1238.

Stokols, D. (1977). Origins and directions of environment-behavior research. In D. Stokols (ed.), *Perspectives on Environment and Behavior.* New York: Plenum Press, pp. 1–36.

Stokols, D. (1978). A typology of crowding experiences. In A. Baum and Y. Epstein (eds), *Human Responses to Crowding*. Hillsdale, New Jersey: Erlbaum.

Stokols, D. (1982). Environmental psychology: A coming of age. In A. Kraut (ed.), *The G. Stanley Hall Lecture Series*, Vol. 2. Washington, D.C.: American Psychological Association.

Stokols, D. and Shumaker, S. A. (1982). The psychological context of residential mobility and well-being. *Journal of Social Issues*, **38**, 149–171.

Tichenor, P. J., Olien, C. N. and Donohue, G. A. (1976). Community control and care of scientific information. *Communication Research*, **3**, 404–424.

Thompson, M. (1983). *To Hell with the Turkeys! A Diatribe Directed at the Pernicious Trepidity of the Current Intellectual Debate on Risk* (Working paper RC-5), College Park, Maryland: Center for Philosophy and Public Policy, University of Maryland.

Torry, W. I. (1979). Antropological studies in hazardous environments: past trends and new horizons. *Current Anthropology*, **20**, 517–540.

Trimble, J. E. (1980). Forced migration: Its impact on shaping coping strategies. In G. Coelho and P. Ahmed (eds), *Uprooting and Development*. New York: Plenum.

Tversky, A. and Kahneman, D. (1981). The framing of decisions and the psychology of choice. *Science*, **211**, 1453–1458.

Vlek, C. (1984). *Large-scale Risk as a Problem of Technological, Psychological and Political Judgment*. (Heymans Bulletin HB—84-691 EX), Groningen, Netherlands: Institute for Experimental Psychology, Social Psychology Division, University of Groningen.

Vlek, C. and Stallen, P. J. (1980). Rational and personal aspects of risk. *Acta Psychologica*, **45**, 273–300.

Vlek, C. and Stallen, P. J. (1981). Judging risks and benefits in the small and in the large. *Organizational Behavior and Human Performance*, **28**, 235–271.

Weinstein, N. D. (1979). Seeking reassuring or threatening information about environmental cancer. *Journal of Behavioral Medicine*, **2**, 125–139.

von Winterfeld, D. and Edwards, W. (1984). Patterns of conflict about risky technologies. *Risk Analysis*, **4**, 55–68.

White, G. F. and Haas, J. E. (1975). *Assessment of Research on Natural Hazards*. Cambridge, Massachusetts: MIT Press.

Willems, E. P. (1977). Behavioral ecology. In D. Stokols (ed.), *Perspectives on Environment and Behavior* (pp. 39–68). New York: Plenum Press.

Witt, W. (1976). Effects of quantification in scientific writing. *Journal of Communication*, **26**, 67–69.

PUBLIC PERCEPTIONS OF SCIENCE: WHAT SEASCALE SAID ABOUT THE BLACK REPORT

S. M. MACGILL

School of Geography, University of Leeds, Leeds LS2 9JT, U.K.

Abstract

The paper investigates the relevance of professional science to a local community in the context of a prominent environmental risk controversy. Specifically, it investigates reaction to the Black Report by the inhabitants of Seascale, the community most closely identified with the Sellafield nuclear risk controversy. The paper draws on an original interview-based social survey among the inhabitants of Seascale. The paper invokes both quantitative and qualitative approaches to the analysis of survey findings. As well as routine examination of response frequencies to multiple choice questions, a comprehensive analysis of open-ended responses is also undertaken. This comprises categorization and interpretation of response types, along with examination of the degree of homogeneity and consistency, or variability and contradiction, in what people say. For the latter, a new methodological tool for reconstructing aggregate patterns of responses (technically, a Galois lattice approach), is deployed. It is found that the Black Report was received in Seascale with a mixture of confusion, satisfaction, deferred judgement and sharp criticism.

Introduction

The Black Report (1984) was the culmination of a scientific inquiry commissioned by the U.K. government in November 1983 to investigate an apparent increased incidence in cancer in the vicinity of British Nuclear Fuels' (BNF) spent fuel reprocessing operations at Sellafield, West Cumbria. The present paper draws on an interview-based household survey undertaken four months after the publication of the Black Report in order to understand the relevance of that report to the community whose situation, in terms of an apparent increased incidence of cancer, was being investigated.

As a case study in the public perception of science, in the context of questions of risk acceptability, the present paper offers contributions both of an empirical and of a methodological kind. Empirically, the paper answers significant though often neglected questions about the relevance of expert scientific inquires to the communities whose situation they primarily address and to whom to a large extent they ostensibly speak. Do such inquiries have a substantial impact on people's outlooks—in the present case, in terms of perceptions of risk—or are they seen merely as ritual acts of political expediency (see Wynne's 1977 interpretation of the Windscale public inquiry), or elite pre-occupation (see O'Riordan, Kemp and Purdue, 1985 on the Sizewell public inquiry)? Such questions as this are of relevance to growing contemporary interest in the public understanding of science (Royal Society, 1985; Layton *et al.*, 1986), and in risk communication and related issues of risk acceptability (Earle and Cvetcovich, 1983; Baram, 1988). In its empirical content, the present paper deepens and extends findings reported elsewhere on the relevance of the Black Report to communities in West Cumbria (Macgill, 1987).

Methodologically, the paper synthesizes quantitative with qualitative approaches to the analysis of interview-based survey findings. In addition to routine examination of response frequencies arising from multiple choice questions, the paper promotes a particularly comprehensive basis for the analysis of open-ended responses, deploying a relatively underutilized discourse-based approach to the elicitation and representation of people's attitudes. This is in keeping with an argument, set out at greater length elsewhere (Macgill and Berkhout, 1986a; Macgill, 1987), to allow for the full richness and idiosyncrasy of people's positions to be represented, *inter alia*, avoiding the artificiality and superficiality of exclusive reliance on pre-determined multiple choice response categories in survey-based research. This is deemed particularly important when, as in the present case, the issue at the focus of the research, Sellafield, is so manifestly a part of local life and normal conversation for the population of interest. Whatever discourses people choose to invoke in responding to questions asked are to be accepted as legitimate measures through which to represent and understand their positions.

As well as appreciation of the popularity of particular types of discourse, a new methodological tool for (re)constructing aggregate patterns of response will be employed. This will be particularly powerful in revealing, on a comprehensive basis, the degree of consistency and firmness in people's positions, and the degree to which there are contradictions and unresolved, perhaps unresolvable, tensions, doubts, ambiguities and uncertainties. The methodological tool to be deployed, entailing examination of what are technically called the Galois connections of the relation between respondents and the remarks they make, is described in the Appendix to this paper. By dealing with the complete set of people's responses, the method avoids the unrepresentativeness that can arise when only a small number of complete quotations is reproduced, insightful though this can be. The method thus combines the comprehensiveness of a quantitative approach with the interpretative insight of a qualitative approach. The paper takes as one of its explicit, though subsidiary, purposes the presentation of a substantial empirical interpretative analysis undertaken on the basis of this method.

The Black Inquiry

Science can take many forms. The Black Report was a document of more than a hundred pages, issued through the Department of Health and Social Security including: a review of epidemiological evidence; a description of some environmental aspects of the Sellafield site, the nuclear power industry in the U.K. and other environmental factors in West Cumbria; a review of radiation exposure of young people in Seascale (with an appendix on radiation and its biological effects); discussion of aspects of risk perception and risk acceptability; and conclusions and recommendations. This was produced by a committee of six experts (two epidemiologists, a medical statistician, a paediatrician, a radiobiologist and a radiation physicist), chaired by Sir Douglas Black. The committee secretary was a medical scientist from the DHSS. Their terms of reference had been 'To look into the recently published claims of an increased incidence of cancer in the vicinity of the Sellafield site: (1) examine the evidence concerning the alleged cluster of cancer cases in the village of Seascale; (2) consider the need for further research; (3) and make recommendations'.

The committee met frequently (usually in London) between November 1983 and July 1984 when its report was published. In the intervening period it received written and, in camera, took oral evidence from representatives of Government agencies and other interested organizations and individuals. Many of these also submitted written reports. The National Radiological Protection Board prepared three specific reports, and other individuals and organizations made available the results of what at the time were unpublished epidemiological studies. These specially-produced scientific studies point to the novelty of the problems faced by the Black Committee. It did not and could not have the strength of a definitive laboratory scientific study, in which measurements can be made in detail and with precision, models can be extensively tested and properly calibrated, and boundary conditions can be specified in a controlled manner. It was instead a case of making the best of imperfect measurements and inferences, and resolving conflicting expert judgement in some way. The eventual publication of the Black Report was followed by controversial debate, in the pages of the press and scientific and medical journals such as the Lancet, New Scientist, The British Medical Journal, and the Guardian.

The Black Inquiry had been prompted by a highly controversial and unusually widely publicized television programme broadcast in November 1983 alleging there to be a connection between discharges of radioactivity from British Nuclear Fuels (BNF) spent-fuel-reprocessing operations at Sellafield in Cumbria (otherwise known as Windscale) and an unusually high incidence of leukaemia among children in the vicinity—ten times the national average for under fifteen year olds in Seascale, the nearest village. Radiation is the only known cause of leukaemia within the limits of present knowledge, and Sellafield discharges greater quantities of radioactivity than any other installation in the U.K.

Government response to the television programme had been unusually swift, with an announcement within twenty-four hours that it had appointed a committee of inquiry under the chairmanship of Sir Douglas Black to investigate matters that had been highlighted. At the time this inquiry was the most significant yet undertaken into possible public health effects of the nuclear power programme in Britain. Its report, published within 9 months of its commissioning, stood as the foremost expert statement on the highly sensitive issue of whether children near Sellafield were environmental victims of nuclear power (Black, 1984). Though tempered with the recognition of various uncertainties and identifying a need (with corresponding recommendations) for more research into the issues that had been investigated, the overall conclusion of the Black Report was that a 'qualified reassurance' could be given to people who are concerned about a possible health hazard in the neighbourhood of Sellafield. The message of reassurance was repeatedly stressed by Sir Douglas in a large number of mass media statements following the publication of the report (Walker and Macgill, 1985).

The Black inquiry can be counted among a growing number of expert scientific tribunals established in order to assess cases of the apparent imposition of serious pollution effects on local populations. Such tribunals are widespread phenomena in the U.S.A., and prominent recent examples in the U.K. include the Committee of Inquiry under Professor Lenihan into morbidity in the Bonnybridge-Denny area of Scotland (Scottish Office 1984), the group investigating an apparent excess of leukaemia in East Yorkshire (ongoing), and (the case considered in the present paper) that under Sir Douglas Black into the possible increased incidence of cancer in West Cumbria (Black, 1984). The purpose of such inquiries can instructively be interpreted as being three-fold. (1) Partly 'scientific'—to 'find out the facts' about local morbidity and mortality effects, often imbued with an expectation of a 'forensic' judgement as to possible causes (and perhaps also confirming or refuting allegations of linkage with locally suspected potential hazard sources); (2) partly to yield guidance to policy makers—for example, recommending that additional controls be enforced on local potential hazard sources, or that particular sorts of research studies be set in train; (3) Partly to assuage public opinion—offering reassurance to those concerned about possible health hazards, or at least pointed to as evidence that matters are being properly looked into.

In an earlier paper (1) consideration was given to the Black Inquiry's position in relation to the first two of these three roles, and of their crucial and ambiguous interactions—the second subtly affecting and distorting the outcome and conduct of the first. In considering the social relevance of the Black inquiry, the present paper will implicitly address its position in relation to the third.

Case Study Research

The recently undertaken interview-based household survey which provides the empirical base for this paper involved a 1 in 14 randomly-selected sample of the adult population of the village of Seascale (135 respondents). Seascale (see Fig. 1) had been designated as a centre of attention in the crucial television documentary due to the relatively high number of cases of leukaemia and other cancers found among its young inhabitants, with its proximity to the Sellafield site giving these particular significance.

Seascale, a village of rather more than 2000 inhabitants, has a very distinctive social composition and history. It is a works village, built in the 1950s by the then Ministry of Supply to house its staff before and at the time that the Windscale piles were under construction. BNF, the present operators at the site (since re-named Sellafield), continue to own a significant proportion of the houses, and almost every family in the village has at least one Sellafield worker. This means not only that the village is almost entirely dependent on BNF for its livelihood, but also that Sellafield is the primary focus of village life. What is particularly significant from the point of view of 'risk perception' orientated social survey work there is that conversation, dialogue and discourse about Sellafield among ordinary people is at the same time an every day part of life, and a rich and well developed part of local social fabric. The roots of concern for possible malign effects stemming from Sellafield's operations, or of active rejection of any basis for concern over issues highlighted in the television programme and investigated by Sir Douglas Black, are very deeply embedded, and typically bound-up with people's different experiences of Sellafield.

Table 1 provides a summary of some key biographical features of the Seascale villagers who were interviewed. The 'place of work' frequencies highlight the high degree of dependence on Sellafield (most of those who work 'at home' are wives of Sellafield employees, and many 'retireds' previously worked there). The 'social class' frequencies indicate a relatively heavy preponderance of higher social classes, reflecting the high level of technical expertise and training required of many staff employed in Sellafield's reprocessing operations. These frequencies suggest also that the Seascale 'public' should not casually be thought of as being an ordinary (or typical) 'lay' public.

Very few people (less than 10% of those approached in the village) declined to participate in the survey. The high response rate can perhaps be interpreted to be a reflection of people's interest and involvement in the subject of the questionnaire; and their perception of the legitimacy and impartiality of the survey (seen by some as an

Location map : Seascale, Sellafield and main towns in Cumbria.

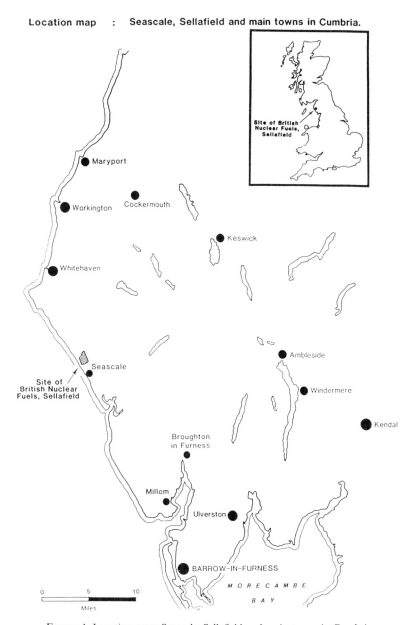

FIGURE 1. Location map: Seascale, Sellafield and main towns in Cumbria.

unusual, possibly unique, opportunity for participating in a public debate which had been very much about their own situation but from which they had been largely excluded). A broad indication of the climate of opinion *vis-à-vis* possible concern for radiation-induced health risks as a result of Sellafield's operations is given in Table 2. It is immediately apparent that Seascale inhabitants are heterogeneous in their positions. Some people speak of concern for radiation-induced health risks as a result of Sellafield's operation (indicating a failure of the Black Report to assuage their concerns—an observation to be elaborated below). Others reject any basis for concern.

TABLE 1

Biographical characteristics of Seascale respondents

| 136 respondents: | 62 male |
| | 74 female |

% by class

| class | 1 | 2 | 3 | 4 | 5 | 6 |
| | 20 | 32 | 18 | 15 | 13 | — |

% by place of work

| Sellafield | Pub admin. etc. | Home | Retired | Other |
| 36 | 12·5 | 23·5 | 15 | 13 |

Seascale Reaction to the Black Report

The survey question

The key (free response) question via which Seascale reaction to the Black Report is to be examined in this paper is:

'What did you think of the presentation and findings of the Black Report?'

This was set within a wider, open-ended questionnaire drawing respondents into a structured conversation about the effect of Sellafield in their lives, their positions in relation to disputed health risks, their views about recent controversial events and activities centred on Sellafield, and their attitudes towards various individuals and organizations actively involved in monitoring, control and lobbying on related fronts. Their responses to the key Black Report question, then, were contextualized within their responses to these other related issues.

In assessing the social relevance of the Black Report below, it is not being assumed that members of the public have necessarily read the report (though some had); the event of its publication and its subsequent dissemination is merely being acknowledged so that, through whatever means (direct, mass media, social network, or whatever), it could potentially have influenced local 'risk attitudes'. It is not within the scope of this paper, nor immediately relevant to the question of the Black Report's social significance, to comment on or seek to remove the various distortions that different means of dissemination will have had.

A selection of complete responses to the key survey question above is reproduced in Table 3. The fullness of some of these responses reflects the awareness of many local

TABLE 2

2a Aggregate Seascale attitudes towards possible radiation induced health risks to children as a result of BNF's operations.

| Not worried | Concerned | Anxious | Cannot say |
| 38·5% | 50·4% | 7·5% | 3·6% |

2b Aggregate Seascale attitudes about beach radiation levels.

| Not worried | Concerned | Anxious | Cannot say |
| 51% | 40·7% | 4·4% | 3·8% |

TABLE 3
Some complete responses to the key Black Report question

1.	In the circumstances, time limited, he did good job. I'm still not happy about it. If there is a bigger risk of leukaemia in this area it might not be Windscale but whatever it is should be found out.
2.	Very technical in a lot of aspects. Very good. Pleased that he didn't state a definite connection with the works and pleased that further research continues.
3.	From all the things we know there's nothing to worry about. Very difficult to prove or disprove. It's like a court case. Black set up as a defence. Given a task with no conclusive answers. Was an accurate and meaningful as he could make it.
4.	Not clear enough—confusing—going round and round the subject. But best they could do in the circumstances.
5.	Accept everything Black said. Waiting for more work. Expected more investigations—never ending. You can always carry on digging.
6.	Black received well in the local area. Inconclusive—it needs another in-depth inquiry. What was done was as much as possible—still left a doubt—we hoped there would be conclusive answer.
7.	Not a lot, didn't have enough time—it hasn't made a difference to local attitudes.
8.	Absolute waste of time. Nothing will be done about it. We've had many reports. Mallory Report about Rugby—nothing has been done—there are still arguments. The powers that be should know what they're doing.
9.	Black was nice enough man. Weren't much good—didn't fill me with confidence. And needs to be looked into further. Didn't give a positive answer. Didn't say BNF is not responsible for these leukaemias. Not direct.
10.	Black, steady, genuine approach—right man for the job. A lot of it was good information. But an air of doubt and started further studies, further research, started the ball rolling. Clusters in different parts of the country—wanted reasons for these—wanted more comparison of environments.
11.	Not readable, could have been clearer presentation—should have been summary report for local people, lay people couldn't make much sense. Poor report, in the course of my work I read many reports, reviewed medical statistics already available, NRPB work more thorough. I would have liked to identify common causes—did all the cases live on farms, near the sea, fathers who worked in the same building. Limited time but no headway.
12.	They ought to have someone looking at it to see what facts are really true.
13.	Taken no interest in what it was saying.
14.	It answered all the questions for me.
15.	Don't think he really settled it one way or the other — it still left you guessing.
16.	Wasn't publicized enough to the rest of the country. People wouldn't have taken any notice—people think there's something terribly wrong up here.

people of the Black Report and their positions of having been drawn into related discussion and debate. The diversity of positions reflects the diversity of local attitudes to the Black Inquiry and the issues by which it was surrounded. The complexity of certain individual responses reflects a state of suspension, uncertainty or doubt about the issues on the part of some people. In order to interpret and represent the complete set of responses as a whole, in a manner that would be faithful to the breadth, popularity and diversity therein, a number of different stages of analysis were undertaken.

Stage 1: Identifying categories of remark
   A detailed schedule of statements. In the first stage of analysis distinct types of statement which could be discriminated within individual responses were identified.

Accordingly, each individual response was divided into constituent statements, and statements (from different individuals) that were deemed to be similar to each other were grouped together in a given category. It cannot be claimed that there was a uniquely correct way of doing this; all that could be done was to discriminate one statement from another according to pre-conceptions of what constituted sensible discrimination of one type of comment from another. The discriminations in this case were partly intuitive, and partly informed by a prior knowledge or awareness of characteristics, general and particular, about the Black Report, the nuclear industry in West Cumbria, media reportage, government reaction, environmental politics and so on. Other analysts may well have discriminated differently: all that could be attempted was to project an interpretation which was considered faithfully to portray the sentiments and attitudes being expressed by the survey respondents, and reflect a familiarity with local conditions and sentiments.

The finest level of resolution that was chosen in discriminating within people's responses produced the 66 different types of statement listed in Table 4. There is a spectrum from 'very favourable' to 'very unfavourable', though the grading between these extremes is by no means simple and linear, as discussed further below.

Almost everyone's response consisted of several of the remarks discriminated in Table 4, so that processing of the complete set of responses in the compilation of this Table, whereby full responses have been dissected into a series of constituent phrases (semantic units), represents a degree of disruption of people's actual utterances; and different parts of what was originally a single response now appear in different parts of the Table. Problems and paradoxes are unavoidable in the process of categorization of a large number of semantic units—see Macgill and Berkhout (1986b) for a fuller appraisal of certain aspects. (In a later stage of analysis, however, a re-construction of how often particular types of statement occurred by themselves, and how often with other statements of various kinds will be given.)

An aggregate schedule of topics. In order to make interpretative analysis more manageable, a reduced schedule of 20 topics (from the previous 66 statements) was

TABLE 4

Black Report statements as discriminated at finest level of resolution

	Topic Code	
General	A	Very good report/Very well done
evaluation	A	I agree with the report's findings
	A	Satisfied/Happy with report/Findings/Report was helpful
	B	Accurate/Detailed/Scientific report
Specific	B	Fair/Unbiased/Genuine report
evaluation	B	Conclusive report/Sellafield is not the cause of cancer
	C	People should look at other causes
	C	Made me aware we need to know the causes of cancer in this area
Effect	D	It set my mind at rest/Reassuring
	D	Reassuring for the community. I'm confident in Douglas Black
Presentation	E	Well presented/Pretty good presentation
	E	Easily understandable/Well written too
		Pleased with the recommendations
	W	Acceptable findings/Adequate report/Mildly reassuring/Sensible
	F	They did the best they could with time and resources they had/Can't determine the cause of cancer so quickly

TABLE 4—*Contd.*

General evaluation	F	All I expected/Much as I expected/Predictable/Not surprised
		He knows more than me/You have to believe the experts
		Happy that something is being done/And that we're being told
		It gave evidence of high incidence of child leukaemia
	G	The numbers were too small to come to a conclusion
	G	The public can't understand the concept of uncertainty in science
Specific evaluation	F	It was done in a hurry/There wasn't enough time
	G	No-one knows everything/There was a lack of evidence
	G	You can't prove a negative—that it isn't caused by Sellafield
		Report showed there's no more risk living here than anywhere else/All comparative risk arguments
	B	I was happy that no connection was made with the factory
	H	Report attracted a lot of controversy
	H	Report was trying to please everybody
Effect	H	Report interpreted differently by different people/groups
	H	The media treated the report badly/In an anti-way
		Report hasn't made any difference to local attitudes
		Report was presented all right/They seemed capable people
Presentation	J	Report could have been put over better/Boring report
	J	Not a very clever presentation/I was not confident about the presentation/Poor presentation
	J	A shortened version would have been useful for the man-in-the-street
		Black was trying to reassure the public (rather than find the truth)
	T	Open for more research/More research needed/This is only beginning
	K	Inconclusive/Not Definite/Inadequate/Left a doubt
General evaluation	K	Vague report/Woolly
	K	Confusing/Didn't come out one way or the other/Said so little/Ambiguous
	V	Rather thick/Too long/Drawn out/Heavy going
	M	It didn't find anything new/Didn't tell us anything new
	L	I didn't think anything about it/I don't know/I've forgotten
Claim of ignorance	L	I can't say/Can't comment
	L	Didn't hear about the findings
	L	Not interested in the findings
	N	Report was a hedge/I am suspicious of the report
	N	I wanted to believe the report but couldn't 'Unreassuring'
General evaluation	N	Disappointed/Not impressed/Unconvincing
	N	Unconvincing for the general public
		Mixed feelings
	P	Figures can be manipulated to say what you want them to
Specific evaluation	V	I couldn't understand the report/Too technical for the layman
	Q	Biased towards BNFL/Report too cautious
	P	Incomprehensive/Limited/Unscientific
	M	Didn't get to the bottom of it/Not in enough depth/Didn't tell enough
Presentation	Q	Disclosure couldn't be too risky/Politically worded report
	Q	He said what they wanted him to say/I think he'd been bought
	R	Not happy with the report/Whitewash/Cover up
General	R	Things are still being covered up in the interests of BNFL
	M	It didn't help anyone by being undecided
	R	I wasn't happy with Douglas Black/He didn't seem to be able to put things very well
	S	Alarming report/Upsetting/Proved we were at risk
Criticism	S	Contradictory report/Conclusions incompatible with the data
		Waste of money/Useless exercise/Load of rubbish

TABLE 5
Alphabetical coding of topics

A	Commendation of report and findings in general terms.	M	Observations that the report didn't get anywhere.
B	More explicit commendation of report's science and conclusiveness.	N	General criticism of the Report's credentials and conclusions.
C	Observations of the need to research into other causes.	P	Specific criticism of Report's science and statistics.
D	Report reassuring to respondent/community.	Q	Critical of the underlying drift of the Report.
E	Commendation of presentation.	R	Critical of Black's deliberate misinformation.
F	Limited time/did the best they could/all I expected	S	Other comments connected by a basically critical attitude.
G	Uncertainties in report science.		
H	Hedged comment about the style of the report.	T	More research is needed.
J	Criticism of the presentation.	V	The Report was difficult to understand for the layman.
K	General criticism of the content— vague, inconclusive, confusing.	W	Report as adequate and acceptable, though of little interest.
L	No comment/not interested/No knowledge.	X	Other statements, not coded as topics.

generated—see Table 5. The reduction involved grouping together similar statements, thereby extending and deepening the process of interpretation begun at the first stage of analysis. For example, 'Acceptable findings' and 'Adequate report' were deemed to be similar to each other, and different from statements such as 'You can't prove a negative' or 'There was a lack of evidence/No one knows everything'. Similarly, 'People should look at other causes' and 'Made me aware we need to know other causes of cancer in the area' would be grouped. Statements uttered with low frequencies were not grouped into topics, because the concern was with a comprehensive overview of core arguments and ideas.

The breadth of meaning for each of the resulting 20 topic categories is not equivalent. Some are more complex and ambiguous than others and appear to have a wider span of meaning. Again, all that could be done was to provide a suitable groupings with a 'reasonable' interpretation of their meanings.

Particularly noteworthy in Table 5 is the relatively large number of categories that might be described as 'unfavourable'. It is to be understood in this connection that the categories in themselves do not constitute a monitoring of opinion—they are not in themselves frequencies or counts—but, rather, a scale of measurement in terms of which 'opinion' can be measured. (Comment on frequencies per category is made later.) It is considered particularly important that the primary scales of measurement—the different categories of topic—had originated from readings of the responses of the people whose positions the analysis sought to reflect and comprehend, and had not been pre-imposed in the analysis.

In the next subsection the meanings understood to be contained within each of the 20 topic categories identified in Table 5 are summarized, elaborating on their additional significance as core lines of argument that people invoke in collecting, rationalizing and justifying their own positions.

A schedule of key topics. Topic A. This is a commendation of the Report and its findings in general terms and an expression of agreement with its conclusion. It is not necessarily an agreement based on the data and scientific method of the Report but an agreement based on more general grounds and often, it turns out, on opinions already held.

Topic B. These are comments about the Report as good, comprehensive science. They might refer to the high quality of data used, the systematic method, the objective reasoning, and the search for truth.

Topic C. These comments acknowledge that there is a local excess of leukaemia, that Sellafield is not the cause, and that there is a need for research into other possible causes.

Topic D. These are comments to the effect that the Report was reassuring to the respondent a community—any doubts which were held during the YTV programme controversy have been assuaged (partially or completely) by the Report. For these respondents the Report achieved what Black had set out to do.

Topic E. These are comments commending the good, clear presentation of evidence, argument and conclusions in the Report, by Sir Douglas Black and other parties involved.

Topic F. These comments recognise that the Black Committee was limited by the time available to prepare the Report. Comparisons might be drawn between this investigation of cancer rates and long-time horizons of cancer research in general.

Topic G. These comments refer to problems of evidence, burden of proof and theorizing in an ambiguous area of science, and how these problems (which have previously only been available to professional science) can be negotiated to a non-scientific lay public.

Topic H. These are comments on the style of the Report and its reception by the media and interest groups. They remark of the Report being written so even-handedly that instead of bringing an end to the arguments it actually generated more argument.

Topic J. These comments criticize the Report's presentation. (Not very clever presentation, Poor presentation, 'Black' did not seem to be able to put things well). For these respondents, the public performance of the Report by its authors and other institutions was seemingly not equal to the performance of the critical view (YTV, Greenpeace, the 'media' in general).

Topic K. These are criticisms of the content of the Report (vague, inconclusive and confusing). The Report has too many qualified statements and conclusions which are not helpful to the advance of the debate, or as an instrument of public policy or to the final meeting of uncertainty with certainty.

Topic L. No comment/Not interested/No knowledge—self-explanatory remarks.

Topic M. These comments remark that the report contained no new evidence or revelations, no new solutions, no new explanations or proofs. There was no persuasive elaboration of the debate as it was understood by the public.

Topic N. These are generalized criticisms of the credentials, goals and persuasiveness of the Black Report's conclusions, expressing disappointment with the report and the manifest problem of legitimation.

Topic P. These points constitute a specific charge that the Black Committee manipulated statistics to prove its own case—in the same way, perhaps, that the producer of the crucial television documentary had manipulated his.

Topic Q. These remarks signify criticism of the underlying drift or style of the Report (its investigation and its text) as too deferential or sympathetic to BNF's case of non-culpability. They relate to Black's spoken intention to 'reassure public opinion'.

Topic R. These are remarks critical of the Report as deliberate, covert misinformation, alleging Black's complicity in allowing it to become another organ of the nuclear industry's own publicity 'whitewash', 'cover-up'—the general perception of the industry as secret, self-interested, and powerful enough to maintain this self-interested secrecy (including the silencing of apparently impartial government inquiries like Black).

Topic S. This topic heading represents a number of different comments connected by a basically critical attitude. (i) Report as 'alarming' (comparable with YTV programme). (ii) The Report contained contradictions/its conclusions didn't match up with the evidence it reported. The report was defective as scientific inquiry.

Topic T. These comments remark that more research is needed; the Black Report is only the beginning. Such remarks have linkages with F. 'They did the best they could'.

Topic V. These are comments to the effect that the Report was difficult to understand for the lay-man. They portray a lay-man's alienation from technical languages—a common theme in sociological work on professionalism and the exclusivity of technical languages.

Topic W. These remarks describe the Report as Adequate and Acceptable: that it was good enough as a piece of public policy, but not very interesting or relevant to the public.

Topic X. Other statements—not coded as topics.

Stage 2: Frequencies, and inter-connections between remarks

Overview. The chosen method in this paper of characterizing people's responses to the science of the Black Report involves not only an appropriate categorization of types of remarks, but also a reflection of (a) their popularity, and (b) their juxtaposition alongside others of various kinds, thus putting individual remarks into perspective by taking account of other remarks by which they were qualified. The popularity of remarks is shown through the (ranked) topic frequencies displayed in Table 6. The symbols $\sqrt{}$, ? and × give a broad indication as to whether remarks are of a broadly favourable, broadly uncertain or a broadly unfavourable kind. A simple count of frequencies—119 favourable, 128 uncertain and 104 unfavourable—again portrays the Black Report as having had something of a mixed reception within Seascale village.

The initial portrayal is very much corroborated in a detailed examination of the aggregate pattern of responses. Here, following the methodology outlined in the Appendix to this paper, the frequencies with which, in people's original responses, particular types of remark were uttered alongside others, is examined (Table 7). This additional stage of analysis would be unimportant if people's individual responses had consisted of similar kinds of remarks. However, it is of considerable importance where, as in the present case, many consisted of different, sometimes seemingly contradictory, types of remarks: a mixture of 'favourable' with 'unfavourable' and 'uncertain', for example. The analysis can only then be sufficiently faithful to people's actual utterances. In presenting this more detailed examination, each of the broad classes of comment will be considered in turn— broadly favourable, broadly uncertain then broadly unfavourable—discussing individual types of topic within each class in turn, beginning with the most popular.

Broadly favourable remarks. Favourable reaction to the Black Report among Seascale inhabitants, as detected by the questionnaire and organized at the derived

TABLE 6

Topic Code	Freq.	Rank	Type	Description
T	46	1	?	More research is needed.
K	36	2	×	General criticism of content of report—vague, inconclusive, confusing.
F	36	2	?	Did the best they could in the time available.
E	35	4	√	Report well presented, both in itself and by the media, i.e. general commendation.
L	30	5	—	No comment/no interest/no knowledge.
D	27	6	√	Report reassuring, both to the respondent and the community.
B	26	7	√	Specific commendation of report; report as good, comprehensive science; conclusive.
V	24	8	×	Report difficult for lay person to understand.
M	23	9	?	Report contained nothing new; didn't get anywhere.
X	23	9	—	(other comments, not coded as topics).
W	22	11	√	Report as adequate and acceptable, though of little interest.
N	14	12	×	Generalised criticism of credentials and conclusions of report.
J	14	12	×	Criticism of presentation.
A	9	14	√	Commendation of report in general terms.
G	8	15	?	Problems of evidence and burdens of proof and their public negotiation.
C	8	15	?	Observations of the need to research into other causes (other than Sellafield).
H	7	17	?	Hedged comment about the style of the report.
Q	7	17	×	Critical of underlying drift of report.
P	4	19	×	Specific criticism of reports science and manipulation of statistics.
S	4	19	×	Other comments connected by a basically critical attitude.
R	1	21	×	Unhappy with Black and his report as deliberate misinformation.

Total number of remarks 404

Total	√	119
Total	×	104
Total	?	128
Total	—	53

level of 'topic' discrimination, falls into five dominant groups, given below in order of their popularity of occurrence:

E The report was well presented
D It was reassuring
B It was fair and accurate
A It was well done and satisfying
W It was a credible government report

The most popular favourable remark is topic E (well presented) which ranks fourth in the schedule of topics overall (Table 6 and line 1 in Table 7). It is a simple and direct response to the first component of the question that was put (presentation). As remarked earlier, presentation is not specifically about the substantive content, findings or recommendations of the report, but about its overall appearance and legitimacy.

Table 7

Abbreviated representation of topic connections*

Level 1	? (T, 46)	× (K, 36)	? (F, 36)	✓ (E, 35)	− (L, 30)	✓ (D, 27)	✓ (B, 26)
	× (V, 24)	? (M, 23)	− (X, 23)	✓ (W, 22)	× (N, 14)	× (J, 14)	✓ (A, 9)
	? (G, 8)	? (C, 8)	? (H, 7)	× (Q, 7)	× (P, 4)	× (S, 4)	

Level 2	(KT,17) (ET,15) (FT,12) (BT,12) (DE,11) (EK,11) (TW,10) (EW,10) (EF,10) (DF,9) (DT,9) (BD,9) (TV,9) (TX,9) (KX,9) (NT,8) (FV,8) (MV,8) (KM,8) (BF,7) (BV,7) (FM,7) (AB,7) (NV,6) (FJ,6) (DW,6) (KV,6) (DV,6) (FK,6) (KN,5) (BW,5) (CT,5) (MT,5) (BQ,5) (VX,5) (FL,5) (VW,5) (JM,5) ...other pairs at lower frequency.

Level 3	(KTX,6) (ETW,5) (DET,5) (EKT,5) (DTV,4) (BDF,4) (KNT,4) (EFT,4) (BKT,4) (KTV,4) (TVW,4) (FTV,4) (DFT,4) (ABV,3) (ETV,3) (BET,3) (BFT,3) (FJM,3) (JMV,3) (KMV,3) (BDQ,3) ... other triples at lower frequency.

Level 4	(EKTW,3) (KNTX,3) (DTVW,3) (DETV,2) (DEKT,2) (ENTW,2) (BDET,2) (BFTV,2) (KMVX,2) (DFTV,2) (BDFG,2) (ABVX,2) (FJMV,2) (GRJX,2)... other groups of four at lower frequency.

Level 5	(EKTVW,2)... other groups of five at frequency of 1.

* (L, 30) denotes that topic L was said by 30 people. (CT, 5) denotes that topics C and T arose together in the responses of 5 people. See Appendix for further guidance in the interpretation of this table.

Of the 35 respondents who made this remark (and reading from level 2, Table 7), 15 qualified it with topic T (open for more research), 10 with topic F (did the best they could) and 11 with topic K (vague, inconclusive). Taking account of the number of people who included two or more of these additional topics in their responses (to avoid double counting) as evident from higher levels in Table 7, over half of the 35 people who said 'well presented' qualified their apparently commendatory remark (topic E) with more uncertain T and/or F, and a further third qualified it with the rather more critical topic K. So, the initial broad labelling of topic E as commendatory needs significant further qualification.

Ranked 6 is the first substantively favourable statement, topic D (report reassuring); proof that the report had achieved what Black had stated intitially was amongst its aims—to reassure local opinion. For these respondents not only had the report been prepared in the proper manner but the results and recommendations had assuaged any doubts which they may have held after November 1983. The report had actually changed their minds, or perhaps more accurately, confirmed to them that there was no reason to be uncertain about local health issues.

Of the 27 respondents who made this remark, 11 qualified it by the further favourable remark E (well presented), 9 with B (good comprehensive science) and 6 with W (acceptable, adequate). These are the highest frequencies with which topic D was complemented (i.e. generally with further favourable remarks). In some cases, however, respondents suggestions of having been reassured were more hedged (see level 3, where D and E appear with T 5 times, and D and V with T 4 times).

Ranked seventh are remarks (topic B) concerning the fairness (the Justice) and accuracy (the Science) of the report. Obviously any document of this type must be assessed according to these criteria, but in this case the significance of these statements is that they were usually uttered in the form of a refutation of the Justice and Science of the earlier YTV programme.

As with D, topic B is qualified mainly by other favourable remarks or remarks which, in keeping with a 'scientifically' couched view, identified uncertainties of a specific kind.

The next positive comment about the content of the report is topic W (adequate report) ranked 11 with a frequency of 22. It is a fairly non-committal, favourable but slightly disinterested general remark about the report and its findings, conveying an impression that the investigation and report were soberly and uncontroversially conducted and written but that in reality, the fact of its being done or its conclusions, little affected local and national debates—it was 'mildly reassuring' and 'sensible'. A fairly distant government report, spoken about in generalities, relevant to (more specialized) argument going on elsewhere.

Comment W occurs most frequently with topics T (open for more research) and E (report well presented, see Table 7 level 2)—a frequency of 10 in each case—each in their turn broadly compatible with the generally positive but non-committal W.

Ranked 14 comes topic A, a collection of more general comments praising the report (a weaker statement of topic B). Topic A has an overall frequency of just 9, 7 of which (see level 2) are respondents who also included topic B (more specifically commendatory) and 3 of these in turn (from level 3) perhaps more paradoxically, included V (difficult to understand) as well.

Broadly uncertain remarks. The group of opinions designated 'Uncertain' can be characterized as consisting mainly of:

T   More research is required to clarify remaining ambiguities in the evidence.
F   The Report contained no more or less than expected, it did not prove to be the advance people might have been hoping for.
M   The report contained little or nothing new.

The top ranked of all comments, given by 46 respondents, is T (open for more research). It seems particularly aimed at the 'qualified' aspect of Black's much repeated message 'qualified reassurance'. Positions remain in a degree of suspension while additional research is undertaken, the time frame for which is unknown. While this view corresponds closely with Black's recommendations for more research, it often refuses to accept as adequate the 'reassurance'. In responses as a whole the remark is accompanied by the criticism that the report was vague (topic K, 17 times), difficult to understand (V, 9 times) and of dubious substantive merit (N, 8 times). However, it is also accompanied by more positive remarks, notably E (well presented) 15 times, B (good comprehensive science—somewhat in contradiction to T) 12 times and W (acceptable enough) 10 times; also by the more hedged 'time limited' comment (topic F 12 times). Overall then, topic T is a remark embedded in many and varied other qualifications.

Topic F ('time limited', 'did the best they could', and 'all I expected'), generally referring to the long time frames of cancer research, was said by 36 people. It is another type of suspended comment, though (from Table 9) can be seen to be more frequently accompanied by favourable or uncertain, than by unfavourable comments.

Ranked 9 is topic M (the report contains nothing new). It typically occurs alongside other uncertain remarks: F ('did the best they could', 7 times) and T (more research is needed, 5 times); and of broadly critical remarks: K (general critisicm of the content of the report, 8 times), V (difficult to understand, 8 times) and J (criticism of presentation, 5 times). It scarcely ever arises alongside commendatory remarks, and as such its broad labelling as 'uncertain' should be further qualified in terms of negative, not positive associations.

Other 'broadly uncertain' remarks were made with lower frequency.

Broadly unfavourable remarks. Statements designated 'unfavourable' are grouped into the following topic categories:

K   vagueness and inconclusivity of report
V   difficult to understand
N   explicit criticism of the report
J   criticism of presentation

Ranked equal second with a frequency of 36 is topic K: general criticism about the confusing and ambiguous nature of the report. The respondents may also, of course, be intimating that ambiguity is often intended not simply read. Nearly half of its occurrences are alongside topic T (open for more research, 17 times). However, it also occurs 11 times alongside topic E (well presented). This apparent contradiction is partly explained by some respondents taking presentation to refer to the way the report had been projected by various media, and not simply the report itself.

Ranked 8 is topic V (remarks to the effect that the report was difficult to understand). This was said by 24 people. These remarks were made alongside a range of others (see Table 7), some apparently contradictory (for example B good comprehensive science; D reassuring; W acceptable, adequate—for some respondents, the report's difficulty seems to have contributed to its overall authority.) Overall then, its initial designation as unfavourable should be qualified in diverse ways.

Ranked 12, said by 14 people, are a range of more explicit criticisms of the report (topic N). Other than being accompanied by each other (which, in this analysis, would not be double counted), they are accompanied typically by other unfavourable, or by uncertain remarks, most frequently (8 times) by topic T (open for more research) and (6 times) by topic V (difficult to understand), and (5 times) by topic J (criticism of presentation). This is much as would be expected for such manifestly critical comments: they virtually never appear with comments designated 'favourable'.

Also ranked 12 (topic J) are criticisms about the report's presentation. Just as 'well presented' is a simple term that might might signify perhaps immediately inexpressible cognitions, so criticism of presentation may signify much deeper rooted doubts. However, its most frequently associated remarks are F (did the best they could, 6 times) and M (nothing new, 5 times). As such it is something of a suspended comment.

Other broadly unfavourable remarks occur with lower frequency, and they will be given no further comment here.

No comment. As well as the range of substantive comments considered above it remains, for completeness, to note the frequency of 'no comment' (30 respondents), and also the existence of a diverse range of statement types (23 in all) each individually of too low a frequency to warrant categorisation into topic categories.

Methodological Evaluation

It would be wrong to privilege the new methodological approach adopted in stage 2 of the analysis above as being uniquely comprehensive or exhaustive. It can simply be said that the scrutiny of inter-connections between remarks has undoubtedly proved to be an effective addition to more familiar statistical procedures for the analysis of survey data. It has enabled systematic consideration to be given to the extent to which individuals respond in a homogeneous, uniform way, and the extent to which they express doubts and contradictions, in giving markedly different types of remark (favourable alongside uncertain and unfavourable). Thus the analysis in this paper reveals two distinct dimensions of diversity in terms of Seascale reaction to the Black Report: diversity between individuals, as seen in people's markedly different positions; diversity within individuals, as seen in the interjuxtaposition of different types of comment. The latter tendencies are of a kind which have perhaps been unduly neglected in the environmental risk perception field as a whole.

A further analytical possibility would be to investigate whether particular significance should be attached to the position of individual remarks within people's full response, with a first comment, for example, taken to be more significant than a second, third or fourth. Having examined a larger data set from this perspective (in fact, the full spatial survey coverage, of which the Seascale responses reported in this paper are but a subset) however, little of consequence seemed to emerge (see Macgill and Berkhout, 1986b). It would also be possible to test for statistically significant correlations between, say, topic categories and respondents' biographical character-istics. This again has been investigated elsewhere (Macgill and Berkhout, 1986b; Macgill, 1988), and is beyond the scope of the present paper. The present paper has been concerned instead to characterize the reaction as a whole to the Black Report that is found among Seascale inhabitants. It is beyond the scope of this paper to attempt a serious review of alternative possible analytical approaches.

It cannot be helped that respondents may not actually have said everything they might have said in responding to the key question analysed above, and on another occasion may respond somewhat differently. Had the above analysis been based on a relatively small number of respondents, or given some absolute status to the frequency response counts in themselves, this may have been quite a serious reservation. What is more important in the analysis as given is to appreciate the richness and diversity of local reaction to the Black Report among the Seascale inhabitants, acknowledging the broad popularity and interjuxtaposition of different arguments.

It is recognised that since some responses are longer than others, they will carry a disproportionate weight in the findings and introduce certain kinds of bias. To seek to rid the analysis of such characteristics would be to subscribe to a view that surveys can achieve a 'democratic' representation of opinion. Although it is sometimes pretended that this can be achieved, it may be truer to suggest that the way in which dominant responses 'bias' the survey analysis reflects the way in which dominant individuals 'lead' local opinion anyway.

It is also worth repeating that the force of the analysis given above does not depend on respondents having read the Black Report at first hand. Its publication is merely taken as an event which occurred, which was projected through various channels (mass media, special meetings, public commentary by certain spokespersons, or more informal social networks) and which, to a greater or lesser extent, could be received by the local population. From the point of view of addressing the research questions posed at the beginning of this paper, to attempt to neutralize the effects of different channels and distortions in the above research would be neither desirable nor possible. Correspondingly, to inquire into ways of lessening the distortion, so as to get the Black

Report's message over more adequately, would demand a different focus from the one actually chosen.

Empirical Evaluation

The Black Report was received in Seascale with a mixture of uninterest, confusion, deferred judgement, satisfaction and criticism. The general attitude polls (Table 2) reflect continued concern among large sections of the Seascale population, though at the same time a significant number of people reject any basis for concern. While many individuals hold firm and steadfast positions, others manifest unresolved doubt and contradictory influences.

The Black Report seems to have been assimilated into an already developed spectrum of risk attitudes in ways which generally seem to have confirmed or deepened people's original positions. It rarely seems to have been a catalyst for revision of position, but was rather, either unimportant to people or, more often, was used to reinforce or extend particular arguments: extending justifications for concern (a hedge, a cover up); extending justifications for the rejection of concern (proving to some that Sellafield was not to blame); or continuing people's doubt and unresolved reflection of contradictory and uncertain influences and arguments—as reflected in the juxtaposition of different types of remarks within many individuals' responses.

For many who harbour continued concerns, the reassurance offered by Sir Douglas Black was clearly wanting in effect. People were not reassured by an externally appointed, albeit independent, specialist offering expert reassurance. This very limited effectiveness of the Black Report in assuaging risk concerns is uncomfortable news for anyone who would want and expect such an outwardly distinguished public policy initiative (and argument of reassurance) to have greater social relevance. Independence of expertise was nowhere nearly enough (cf. the argument of the House of Commons Environment Committee 1986 for experts independent of the nuclear industry to reassure the public); and a procedural mechanism—the science of an expert tribunal— removed from the ordinary experience of everyday life was again largely impotent for handling questions of societal risk acceptability (cf. O'Riordan et al.'s 1985 broadly analogous comment in respect of the Sizewell public inquiry).

The limited effectiveness of the message of reassurance emerging from Sir Douglas Black's Inquiry in assuaging people's concern for possible health risks from Sellafield can perhaps be more closely considered by discriminating between two (linked) aspects—the social and the scientific.

In social terms, limitations arose due to stronger forces acting at interpretational, psychological, cultural, institutional and experiential levels for the Seascale population. Some of these sorts of influences have already been identified in other contexts in the risk perception research literature (Thompson, 1980; Otway and Thomas, 1982; Wynne, 1983; Renn, 1985; Douglas, 1986). An important implication is that the establishment of trust between, on the one hand, concerned populations and, on the other, institutions in a position to offer reassurance, cannot be secured by a one-off initiative by externally appointed specialists offering expert reassurance. It would be necessary to engage much more adequately with the deeply rooted vectors of interest and human reality that are found in Seascale. Given the contrasting and often contradictory risk positions there, any goal of complete reassurance is perhaps one of impossible magnitude.

To these 'social' limitations can be added those of a second kind, by acknowledging that a significant number of Seascale inhabitants recognized many points of entry to dispute the integrity of the science contained in the Black Report—because of immaturities, imperfections and uncertainties in the science of low level radiobiological health effects; and the long time horizons over which cancer research operates.

Conclusion

The analysis of interview-based data from the population of Seascale in the wake of the publication of the Black Report has revealed a rich complexion of risk positions. At a general 'attitude poll' level (Table 2) the manifest diversity of positions is readily apparent. At a more detailed level, it is reflected in the different kinds of arguments, often inconclusive, that people draw on when being invited to talk about their positions in relation to the Black Report. These arguments have been systematically examined and reconstructed though successive stages of analysis in this paper. Substantively, they amount to a formidable web of agreement, dissent, resistence and reflection which cannot easily be changed, shaped or broken down.

Postscript

Finally to place the above analysis in a wider setting of political economy, it can be observed that the Black Inquiry was established as a political response to great concern evoked by a television documentary. This programme was a threat to national confidence in the nuclear industry, and a challenge to the basis of official discharge authorizations. In commissioning Sir Douglas Black's Inquiry, government was responding to these sorts of national issues. Government did not engage with the people who lived in the targeted area—Seascale, or West Cumbria more generally—in setting up the Inquiry. It did not respond directly to concern there, for positions there were unknown. Neither did the Inquiry's report engage very fully with local people. They remained on the periphery to a centre with more powerful interests and concerns. If this interpretation of the political economy of the Inquiry portrays it as a somewhat hollow expedient, then further qualification is called for. For significant new medical and scientific studies were recommended as a result of Sir Douglas Black's Inquiry, and have now begun to produce crucial new findings (for example Gardner et al., 1987). There has in turn been an apparent shift in position on the part of Sir Douglas Black himself.

Acknowledgements

To Frans Berkhout for his work as research assistant during the survey period, and in compiling the statement and topic categories. To Trevor Springer for developing the computer algorithm. The research reported in this paper was partly supported by grants (nos D00232055 and D0025011) from the Economic and Social Research Council.

Reference Note

1. Macgill, S. M., Ravetz, J. R. and Funtowicz. Scientific reassurance as public policy: the logic of the Black Report. Working paper no. 448 (1985).

References

Atkin, R. H. (1974). *Mathematical Structure in Human Affairs*. Heinemann, London.

Black, D. (1984). *Investigation of Possible Increased Incidence of Cancer in West Cumbria. Report of an Advisory Group*, HMSO, London.

Earle, T. C. and Cretkovich, G. (1983). *Risk Judgement and the Communication of Hazard Information: Toward a New Look in the study of risk perception*. Batelle Human Affairs Research Centers, Seattle, Washington.

Gardner, M. J. *et al.* (1987). Follow-up of study of children born to mothers resident in Seascale, West Cumbria (both cohort). *British Medical Journal*, **295**, 822–827.

Ho, Y. S. (1982). The structure of verbal descriptions. *Environment and Planning B*, **9**, 397–420.

House of Commons Environment Committee. (1986). *Radioactive Waste*. Session 1985–86, First Report, House of Commons Paper 253; i-xvii HMSO London.

Layton, D. C. et al. (1986). Science for specific social purposes (MSP); perspectives on adult scientific literacy. *Studies in Science Education*, **13**, 27–52.

Lee, T. R. (1984). Perceptions of and attitudes towards risk. Paper presented to 'Technological Risk' conference. University of Manchester, July.

Macgill, S. M. (1985). Structural analysis of social data: a guide to Ho's Galois lattice approach and a partial respecification of Q-analysis, *Environmental and Planning A*, **17**, 1089–1109.

Macgill, S. M. (1987). *The Politics of Anxiety: Sellafield's Cancer-like Controversy*. Pion, London.

Macgill, S. M. (1989). Risk perception and 'the public's insights from research around'. In J. Brown (ed.) *Environmental Threats*, Pinter (forthcoming).

Macgill, S. M. and Berkhout, F. G. (1986a). *Understanding 'Risk Perception': Conceptual Foundations for Survey-based Research*, WP 453. School of Geography, University of Leeds.

Macgill, S. M. and Berkhout, F. G. (1986b). *Child Leukaemia around Sellafield: Local Community Reaction to the Black Report*, WP350. School of Geography, University of Leeds.

O'Riordan, T., Kemp, R. and Purdue, M. (1985). How the Sizewell B Inquiry is grappling with the concept of acceptable risk. *Journal of Environmental Psychology*, **5**, 69–85.

Otway, H. and Thomas, K. (1982). Reflections on risk perception and policy. *Risk Analysis*, **2**, 147–159.

Renn, O. (1985). Risk perception—systematic review of concepts and research results, Risk Analysis seminar, IIASA. Laxenburg, Austria, November.

Royal Society (1983). *The Assessment of Risk*. A study group report, Royal Society, London.

Royal Society (1985). *The Public Understanding of Science*. Royal Society, London.

Scottish Office (1984). *Bonnybridge-Denny Morbidity Review. Report of Independent Review Group*, Scottish Home and Health Department, New St Andrew's House, Edinburgh.

Slovic, P. *et al.* (1980). Facts and fears: understanding perceived risk in *Societal Risk Assessment*. R. C. Schwing and W. A. Albers (Plenum, New York), pp. 187–213.

Thompson, M. 1980. Political culture: an introduction, IIASA, WP–80–175, Laxenburg, Austria.

Walker, G. P. and Macgill, S. M. (1985). *The Black Inquiry and the Media*, WP 449. School of Geography, University of Leeds.

Wynne, B. (1982). *Rationality and Ritual*. British Association for the Advancement of Science, St. Giles, Chalfort.

Wynne, B. (1983). Public perception of risk—interpreting the 'objective versus perceived risk' dichotomy, IIASA, WP--85–117. Laxenburg, Austria.

Appendix

The derivation and significance of Galois connections

The paper utilises a novel methodological perspective for reconstructing the aggregate patterns of a large number of multiple character responses to a particular survey question, by tracing the frequencies with which particular categories of remark were associated with others. The methodology has its origins in the mathematical theory of Galois connections (Ho, 1982; see Macgill, 1985 for a simplified account) and is used in the paper as a preferred alternative tool to the use of 'nearest neighbour' or 'smallest space' methods which might alternatively have been adopted for the processing and interpretation of multidimensional categories of risk survey data (Slovic et al., 1980; Earl and Cretkovich, 1983; Lee, 1984). The latter are not favoured in the present paper due to their characteristic of transforming, over-generalising, and partially masking actual survey responses in producing interpretative categories. The chosen approach does not have these qualities. As the approach is understood to be new to the risk perception research field, it is outlined in this appendix.

Given 136 respondents and 20 categories of response, the set of responses as a whole can be represented as a 136 × 20 binary rectangular array. Its general form would be the same as that of the hypothetical 6 × 7 array in Table A1, where a 1 in the ith row and jth column represents the fact that the jth topic arose within the response of the ith person (the content of the ith person's complete response can be deduced from the positions of the 1's in the jth column). For convenience, this hypothetical array will be the basis for the following outline of the methodology.

TABLE A1

Topics \ People	1	2	3	4	5	6	7
a	0	1	1	0	0	1	0
b	1	1	1	0	0	0	1
c	1	1	0	0	1	0	1
d	1	0	0	1	1	0	0
e	1	0	0	1	0	0	0
f	0	0	1	0	0	0	0

Reconstructing the frequency with which particular topics occurred with others entails calculating the frequency of occurrence of all possible pairs of topics (i.e. how many people said a and b, a and c, a and d, etc. for all other pairs), calculating the frequency of occurrence of all possible triples of topics (i.e. how many people said a and b and c, a and b and d, etc. for all possible triples of topics), and, correspondingly, calculating the frequency of occurrence of all possible groups of four, five, six and seven topics. (Seven was a set upper limit on the number of topics allowed per respondent in processing the data; hardly anyone said more than this.)

The result of calculating all such frequencies would constitute a definitive reconstruction of all original topic conjunctions and associations, in as much as such a reconstruction can be realized on the basis of the topic categorisation (7 topics in this hypothetical data: 20 for the actual survey question). They are given in Table A2. It can be readily seen that some components in Table A2 are wholly included within others, and in that sense are redundant. They are marked with an 'X'; the information they depict is fully contained in other components, and can therefore be deleted from any further analysis. In fact, 8 of the 19 components can be deleted in this way—a considerable reduction. Technically, what would then be produced could be called the

TABLE A2A
Indicating which topics were said by whom

(a, 236)	(d, 145)
(b, 1237)	(e, 14) ×
(c, 1257)	(f, 3) ×

TABLE A2B
Pairs of topics, and the individual's who said them

(ab, 23)	(be, 1) ×
(ac, 2) ×	(bf, 3) ×
(af, 3) ×	(cd, 15)
(bc, 127)	(ce, 1) ×
(bd, 1) ×	(de, 14)

TABLE A2c
Triples of topics, and the individual's who said them

(abc, 2)
(abf, 3)

TABLE A2D
A four-topic individual response

(bcde, 1)

Galois connections of the relation represented in the original binary array. For the array given in Table A1 they are arranged in the form of the lattice given in Figure A1. Reading successive levels from the foot of the lattice upwards runs as follows.

Level 1—this identifies each of the topics discernible at that level. The total frequency of occurrence of each topic which respondents included in their responses follows simply from a count of the respective number of people: 3 for topic a, 4 for b, and so on. (The reason why topics e and f are omitted from this level is due to their redundancy, as explained above.)

Level 2—this identifies which respondents included each of the given pairs of topics in their responses. The total frequency of occurrence of each pair of topics follows simply from a count of the number of people for each respective pair: 2 people for the pair a, b; 3 people for the pair b, c; and so on. Only those pairs of topics which actually arose within responses are included in the lattice. No useful purpose would be served by cluttering the lattice with fictitious pairs: pairs that did not in reality arise.

Level 3—this identifies which respondents included each of the given triples of topics in their responses. The total frequency of occurrence of each triple of topics follows, as before, simply from a count of the number of people for each triple. As before, fictitious triples are not included in the lattice.

Level 4—this identifies a respondent who uttered four topics within his or her response.

The lattice (more specifically, the Galois connections out of which it is constructed) constitutes a definitive isolation and organisation of all the topic associations and

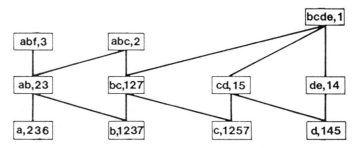

FIGURE A1. Lattice derived from hypothetical data, table A1.

conjunctions from the original set of question responses. It gives a holistic view of topic popularity (frequency of occurrence), and associativity. As we ascend the lattice individual topics are set into relief in terms of other topics through which, to varying extents, respondents extended, elaborated or qualified particular phrases or semantic utterances, to produce their complete responses to the question.

Pragmatically the Galois lattice procedure is more efficient than a more conventional, routine count of frequencies of topics, topic pairings, topic triples and so on, both because of its neglect of redundant components (as already noted) and because of the possibility (via an underlying mathematical theory) of devising an efficient computational algorithm for deriving the Galois connections of a binary relation, with minimum time spent searching through 'fictitious' possibilities. Conceptually, the Galois approach is attractive through its depiction of a comprehensive pattern of dialogue: individual components are seen to be part of a holistic structure, deriving significance from their place within the structure, and not as isolated entities. As a methodology, its advantage over nearest neighbour or smallest space methods lies in its adherence to topic categories rooted in actual utterances, not in interpretative categories further removed. Any particular topic can, moreover, immediately be linked to the individuals within whose response it arose. The approach though, demands more of the analyst in terms of painstaking combing in the case of a data set of any size.

LEVELS OF ENVIRONMENTAL DECISIONS

OLA SVENSON* and BARUCH FISCHHOFF†*

* *Department of Psychology, University of Stockholm, Stockholm, Sweden S–10691*
and † Decision Research, 1201 Oak, Eugene, Oregon 97401, U.S.A.

Abstract

The language of decision theory is used to model the perspectives of two parties in the management of an environmental hazard. These are (a) individuals whose homes are polluted by radon by-products which thereby pose an uncertain health hazard and (b) the public authorities concerned about those individuals' welfare. The analysis provides a way of anticipating ways in which the perceptions and decisions of these parties are intertwined, as well as how they can come into conflict. It also suggests ways in which these difficulties might be ameliorated by altering the respective decision problems. The heuristic value of modeling decision problems in other contexts is discussed, with brief consideration of the substantive issues arising with acid rain and seatbelt usage.

Introduction

Environmental psychology often contributes to the creation of environmental disputes, by documenting the effects of various stressors on people's health and well-being. Investigators have shown how noise interferes with the ability to learn, how workplace pressures diminish the quality of non-working life, and how crowding strains the fabric of interpersonal relations. In some cases, these studies are the original cause of alarm, suggesting effects that people had not even suspected. In other cases, the studies confirm intuitive suspicions, thereby emboldening people to protect themselves, helping them to pick the most consequential issues, and providing ammunition for their struggles.

Environmental psychology can also help to focus and ameliorate disputes. Studies like those described above can have this effect when they demonstrate the existence of effects so incontrovertibly that they cannot be ignored, or their non-existence so convincingly that interest in them flags. Given the weakness of our research methods, relative to the complexity of environmental systems, such strong evidence is seldom forthcoming. As a result, the healing powers of environmental psychology are found more often in attempts to describe the underlying causes of disputes, by characterizing the positions of their participants.

These efforts may take a variety of forms and have a variety of effects. Where fundamentally different ideologies motivate the stances of opposing parties, then descriptions of their world views can help to make their positions more predictable and comprehensible (although not necessarily more credible) (Cotgrove, 1982). Where participants come from different cultures (within the same national society), then a description of the inconsistent ways in which they use terms can improve communications between them (Green, 1982; Slovic *et al.*, 1979). Where the dispute-resolution process (e.g. hearings, referenda) discourages candor and encourages posturing, then a description of the parties' actual beliefs and concerns can clarify what actually motivates them (Fischhoff *et al.*, 1983; Marcus *et al.*, 1984);

a description of the process itself can show how it reflects the values of those who control it (Kemp *et al.*, 1984). Where experts question laypeople's ability to participate in environmental management, then descriptive studies can clarify how much the latter know, how much they are capable of learning and to what extent their beliefs are clouded by emotion (Fischhoff *et al.*, 1981; Gricar and Baratta, 1983; Vlek and Stallen, 1980). Conversely, studies of experts' judgments can show the definitiveness of the advice that they offer (Fischhoff, 1985; Murphy and Winkler, 1985).

One strength of all these techniques is that they are quite open-ended, in the sense of being able to reveal diverse patterns of behavior. That strength brings a certain weakness, in that investigations are initially unstructured, relying on the creativity and sensitivity of the investigators to discern patterns. A second strength is that they are non-prescriptive, in the sense of offering no opinion as to how disputes should be resolved. Their corresponding weakness is that they offer little help in going from understanding conflicts to resolving them. We offer here a complementary technique that builds on the strengths of these techniques, while offering some opportunity to overcome their weaknesses.

Its premise is that people choose their responses to environmental issues through some sort of decision-making process. That decision could concern a personal action (e.g. whether to wear a seatbelt), a collective action (e.g. how a country should respond to pollutants coming from a neighbor) or the link between personal and collective action (e.g. whether to sign a petition regarding an environmental issue). The key elements of those decisions are the key elements of any decision: a set of alternative courses of action, a set of consequences that might arise from taking those actions and a set of beliefs regarding the likelihood of each consequence arising from each action. If it were possible to capture the decision-making problems that the parties to an environmental dispute saw themselves as facing, then one should have a fairly succinct representation of the things that matter to those individuals.

By contrasting the decision problems of different parties, one should be able to get some insight into the reasons for disputes (or the lack of them). In some cases, this will show the disagreements to be deeper than imagined; in other cases, they will appear more shallow or restricted to issues that can be resolved by scientific research or existing political processes. In any case, the air should be cleared and the debate focussed.

The most convenient language for describing decisions is that of decision theory, which was developed by philosophers to describe how rational decisions should be made, by economists to describe how actual decisions are made (under the assumption that all decisions are made rationally) and by various consultants who help real people achieve the economists' ideal (Abelson, in press; Fischhoff *et al.* 1985; Fishburn, 1982; Slovic *et al.* 1985). The theory provides a highly flexible language, capable of describing almost any deliberative choice among alternatives in terms of well-characterized concepts. That description would be consistent with rational decision making, but would not assume it. Indeed, rationality could be used as a point of departure, with decision theory's language being used to describe behavior as deviations from the ideal, perhaps suggesting ways in which to help people be more rational (Fischhoff and Beyth-Marom, 1983; Ungson and Braunstein, 1982).

As a tool for illuminating environmental disputes, the language of decision theory can be used entirely descriptively. By providing readily compared characterizations of different parties' positions, it allows diagnosis of the precise sources of disagreement. Such a diagnosis alone can have salutary effects. It may, for example, heighten mutual respect by tracing conflicts to terminological differences or legitimate conflicts of interest (rather than, say, to stupidity, arrogance, or venality). It can ensure that attention is paid to those (potentially overlooked) areas in which the combatants do agree (Edwards, 1980; Gardiner and Edwards, 1975). More ambitiously, it can focus discussion on the key issues, whose resolution might allow the parties to agree on what action to take (even though they maintain different beliefs and values on many subsidiary issues). Its structuring of the parties' perspectives should explain their response to existing alternatives and may even prompt the creation of attractive new options (Chen, 1980; Hammond and Adelman, 1976). Finally, use of a common nomenclature should help investigators compare results across studies (seeing, for example, if a particular deviation from optimality frequently afflicts hazard managers or the public relying on them).

Neither the language of decision theory nor the attempt to explain the bases of disputes is particularly novel. The main contribution of the present proposal is the combination of the two and the use of decision theory without a presumption of (or particular interest in) optimality. The approach is demonstrated here in the context of a case study dealing with an environmental issue of increasing interest, the accumulation of radon in homes, which illustrates a recurrent kind of dispute, that arising from a conflict between the decision problem faced by individual citizens looking out for their own best interests and that faced by the hazard managers looking out for society's best interests. The descriptions of these decisions are derived by analysing both the scientific facts of the problem and the ways in which those facts emerge in the lives of the different parties. These parallel analyses are guided by the particulars of this issue and a general understanding of how people and policy makers make decisions under conditions of uncertainty.

The Issue

Radium is a radioactive element which decays into radon, a gas, which decays, in turn, into radon daughters. When inhaled, these increase the risk of lung cancer. Recently, scientists have come to realize the magnitude of radon releases into homes from both the underlying ground (e.g. granitic rock and shale) and common building materials (e.g. bricks composed of light-weight concrete). Because both sources release radon at a roughly constant rate, the amount present at any time (and the attendant risk) depends upon how well the structure is ventilated. As a result, all other things being equal, well-insulated homes are riskier than drafty ones, meaning that prudent residents are penalized for listening to the experts' advice regarding energy conservation. Under plausible assumptions (regarding inhalation speed, exposure time, etc.), an average radon daughter concentration of 40 Becquerels per cubic meter of air (Bq/m^3) in a home produces a risk of approximately 1 additional case of lung cancer per 10,000 persons exposed per year. Because radiation risks in this stage are roughly proportional to exposure, a concentration of 400 Bq/m^3 would produce an expectation of 10 additional cases per 10,000 persons exposed per year (Statens Strålskyddsinstitut, 1982).

Much of Sweden faces the unhappy combination of tightly insulated homes built on granitic bedrock, using radium-rich building materials. Today, some 40,000 of

Sweden's 3,500,000 homes are estimated to have radon levels above 400 Bq/m³. Increasing the exposure from 40 to 400 Bq/m³ increases the lung cancer rate caused by this hazard from 0·6% to 6% for the exposed population. This translates to a 300 day decrease in life expectancy. By contrast, the risk of lung cancer is less than 1% for non-smokers and about 10% for smokers. The population dose would be reduced by 15% if all homes had less than 400 Bq/m³, by 40% if all homes had less than 100 Bq/m³ (Statens Stålskyddsinstitut, 1982).

The authorities' decision problem is to determine what constitutes an acceptable dose level for all homes. The residents' decision problem is deciding what to do when faced with a particular exposure. These problems are represented schematically in the *decision trees* of Figures 1 and 2. Reading from the left, these representations begin with a *decision node* presenting alternative courses of action. These actions lead to *event nodes*, showing events that affect the consequences following from pursuit of these actions. These events may or may not be predictable, leading to decision making under conditions of certainty or uncertainty, respectively. Further decision (and event) nodes may arise, until eventually a set of consequences is considered. At times, such qualitative structuring of a problem provides all the insight that decision theory has to offer (as a guide to either making decisions or describing them). At times, there is further benefit to estimating the model's parameters, namely, the *attractiveness* of the consequences of the various action–event sequences and the *probabilities* of their arising.

The Residents' Decision Problem

Although radon poses a threat to most Swedish homes, it is greatest for those built between 1950 and 1975, when construction relied heavily on lightweight concrete. More modern houses not only use safer materials, but must meet a standard of 70 Bq/m³. Older homes were built with non-radioactive materials (and face only the threat from bedrock radiation). For the sake of simplicity, we will focus on the residents of middle-aged homes and assume that all are owners (which is generally true for single-family homes, but not for apartments).

Insofar as there is no radiation standard for these homes, their occupants' initial decision is whether to consider the problem at all. Figure 1 represents this decision as a choice between the residents having their homes tested and doing nothing. A governmental agency conducts the tests at little cost. The event that follows is the existence of a particular radiation level, which is made explicit for those who take the test and is a matter of guesswork for others. The likelihood of these different levels in a randomly selected house from this population is the same in either case (and is discussed in the context of the authorities' problem). What that risk is perceived to be by testers and non-testers would require empirical study (Lichtenstein *et al.*, 1978; Weinstein, 1980).

Once the time of testing has passed, residents face a second decision. In it, they must choose between leaving their house as is, improving its ventilation, drastically remodeling it (to remove the offending building materials), and leaving it for a new home (with a low radiation level). There seem to be four consequences with major significance for one or more of these action–event–action sequences.

(a) *Radiation health risk.* The risk is low either when the house has little

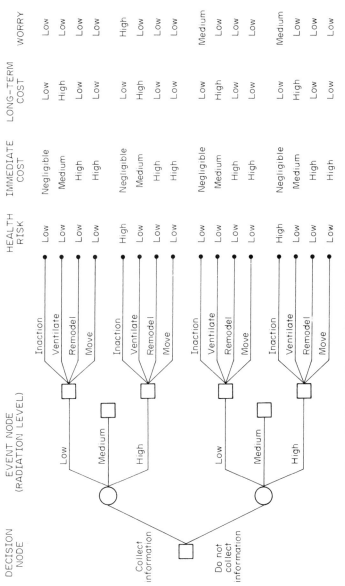

FIGURE 1. The radiation hazard in homes from the residents' perspective.

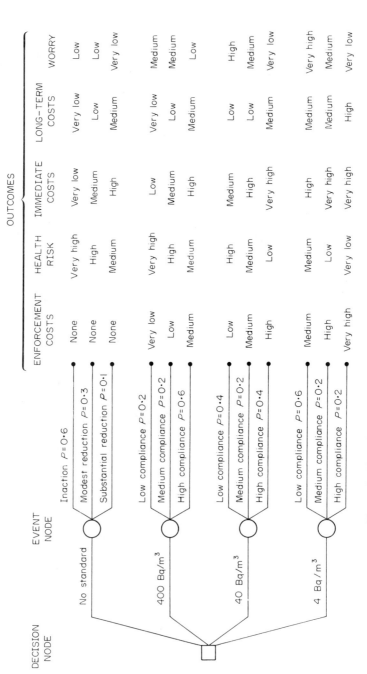

FIGURE 2. The radiation hazard in homes from the authorities perspective.

risk in the first place or when steps are taken to reduce it. It is high only when the resident is inactive in the face of great exposure (known or unknown).

(b) *Immediate monetary costs*. These are negligible with inaction, low with improved ventilation (arising mainly from the need to bolster the heating system), and high with remodeling or moving.

(c) *Long-term costs*. These are substantial only with improved ventilation, due to the cost of heating a deliberately drafty house (Note 1). New and remodeled houses avoid exposure, rather than reducing it.

(d) *Worry*. The risk has received enough publicity in Sweden to make the average resident moderately worried in the absence of information about their specific home (Sjöberg and Jansson, 1982). Worry can be reduced by active steps (of ventilating, remodeling, and moving) or receipt of a low test result. It can be increased by a bad test result, especially when no action is taken.

These evaluations are roughly characterized in the outcome columns of Figure 1. Because all are described negatively, the best possible outcome profile is 'low, low, low, low'. The only way to get that outcome is to do nothing after having been found to have little radiation exposure, making inaction the *dominant* option following a low test result (a conclusion that was obvious without creating a decision tree).

A high test result creates a more complicated situation, necessitating tradeoffs among the consequences. Here, the decision tree proves more useful for clarifying the problem. Inaction leads to high risk and high worry, two aversive consequences that can only be reduced by an active response incurring substantial financial costs. Because the choice among the active responses is purely a matter of economics, the authorities could help residents considerably by providing them with the sort of precise cost estimates that it is difficult (and inefficient) for individuals to produce on their own. If these economic analyses do not point to a preferred action, then it may be useful to enrich the consequence set with factors such as the disruption caused by the options or the regret likely if they do not work as planned. Integrating these factors with the economic factors, and comparing them all with the benefits of reducing the health risk (and its attendant worry), requires evaluative judgments. Although these judgments could legitimately be different for all residents, everyone could be helped by a common set of elicitation techniques designed to help people understand their own values as applied to this issue. Developing such aids to decision making is another service that the authorities could render.

For residents whose houses have tested high, these tradeoffs are difficult but clear cut. By contrast, residents who have avoided testing can only speculate about the risks they face. Any action that they take risks incurring a large expenditure without any assured reduction in risk (but a certain reduction in worry). Although this suggests the advisability for all residents of being tested, the decision tree points to a few exceptions. One arises with the absence of uncertainty; residents who feel that they know their risk even without testing can proceed to the action decision. A second exception arises in the absence of opportunities for uncertainty reduction, as might be perceived by residents who distrusted the scientists and their tests, for whatever reason (Cotgrove, 1982). A third exception arises when residents lack the resources for coping with demonstrated risks. Comparing the respective 'high exposure–inaction' rows with and without testing shows that testing increases worry from medium to high. As a result, the decision to test becomes a choice between a sure thing (the current medium level of worry) and a gamble offering some probability of low, medium and high worry (depending upon the outcome of the test). If it is extremely aversive to know for certain that one is in daily danger but too poor to do anything about it, then it may be quite reasonable to avoid testing.

If this last situation seems unacceptable, then there may be pressure to change the structure of the decision problem. One possibility is to create the new option of having someone other than residents absorb the costs of making their homes acceptably safe. That someone could be the original construction companies (following principles of strict liability) or some branch of government. With this new decision problem, other event nodes and consequences may become pertinent, leading to a redrafting of the decision tree. For example, a key event might be the number of other poor residents who have tested high; if the number is large enough, then they may become an effective lobby for compensation. Indeed, further analysis might show the advisability of refusing the test, so as to help foil the voluntary testing program, so as to force compulsory testing, so as to create a large pool of angry poor residents. Although decision theory cannot create such options, it can prompt the search for them, by showing unacceptable situations in stark relief, and so help to organize thinking about the alternatives.

The Authorities' Decision Problem

As with other environmental decisions, the 'authorities' concerned with radon are found in a number of agencies, subject to the internal and external politics of the allocation of resources within and across institutions. For the sake of simplicity, they are treated here as corporate bodies whose sole criterion when evaluating decision options is the 'public good'. Figure 2 shows a decision tree in which the options considered are alternative standards for acceptable residential exposures. These are: (a) no standard at all, (b) 400 Bq/m^3 (associated with a lifetime cancer risk of 6 in 100), (c) 40 Bq/m^3 (the current standard for exposure in Swedish work environments) and (d) 4 Bq/m^3 (which would make homes 10 times safer than workplaces). Alternative (d) represents a relationship that characterizes the allowable exposures for the public and workers in many other domains—unlike radon, for which homes are currently allowed to be much riskier than workplaces (Derr *et al.*, 1983). The standard for new homes is 70 Bq/m^3, whereas for existing homes there is no firm standard.

The key event determining the consequences of adopting each alternative standard is the degree of public compliance with it. That degree is described here in terms of three representative values: high, medium and low. The probability of achieving each of these rates should depend on both the degree of enforcement and the stringency of the standard (increasing with the former, decreasing with the latter). As a result, higher compliance is associated (in the tree) with higher enforcement costs and lower compliance costs. In so far as one cannot speak of compliance with 'no standard', the event following that action is the level of voluntary action by residents.

The specific probabilities given in Figure 2 are very rough estimates, based on the assumption that residents will do relatively little in the absence of firm action by the authorities (reflecting both the expense of action and the difficulty of knowing what to do). More precise estimates might be derived by detailed analyses extending

Figure 1 to describe the decision problems faced by individual residents. That analysis could rely either on available economic and risk statistics or upon psychological measures of the residents' perceptions (Kunreuther et al., 1978).

At first glance, the consequences of the authorities' actions resemble those of the residents, with the outstanding difference of 'enforcement costs'. These would include both the budget of the regulatory agency and social costs, such as public resentment or confusion within the construction industry. From them, would be subtracted any social benefits of the regulatory process, such as building confidence in government or educating the public about the risks (Fischhoff, 1984).

On closer examination, the similarity of the remaining consequences to their counterparts in Figure 1 proves superficial. One difference is that here the costs and benefits are aggregated across all members of society. Thus, for example, although an individual resident might realize a reduction in risk by leaving a high-radon house, that change would not be counted as a societal gain unless the house stood empty (so that the exposure was not shifted to someone else).

A second difference in the two sets of consequences is their specificity. Although the event nodes specify the proportion of residents taking some action, they do not indicate what those actions will be. In order to proceed, we assume that residents who act are equally likely to move, ventilate, and remodel. Analysing the decisions of individual residents (Figure 1) would be one way of improving on these estimates for Figure 2.

A final difference between the consequence sets is the source of the values that they express. Whereas individual residents have sovereignty over how they weigh the consequences of personal decisions, the authorities are expected to express (and protect) society's values. Thus, for example, although residents may choose ventilation over remodeling because the short-term costs are lower (even though the long-term costs are higher), the authorities should maintain a longer planning horizon when weighing the relative importance of these two consequences. Whereas individual residents' decisions regarding how much to pay for health protection is likely to reflect their personal ability to pay, the authorities should treat all citizens as equally valuable (Note 2).

Assessing the approximate magnitude of the consequences associated with each action–event sequence is relatively straightforward, except for 'worry'. Pending empirical studies of what governs worry in these particular situations (Baum et al., 1983; Elliot and Eisdorfer, 1982), we have assumed that people interpret the stringency of government action as indicating the severity of the radon problem. Therefore, inaction causes the least worry when there is no standard and the most worry when there is a stringent standard. If the costs of worry seem great, then the authorities might hesitate before issuing a standard for which they could not achieve substantial compliance. We have also assumed that compliance with a tight standard eliminates worry, whereas compliance with a weak standard still leaves people uneasy.

Contrasting the Problems

Neither of these decision trees is definitive. In order to derive firm recommendations from them, one would want to flesh them out in various ways, such as assessing the probabilities more precisely, quantifying the consequences, elaborating the event nodes. However, even a sketchy analysis can clarify the structure of a decision problem. In this case, it shows the differences between the options, consequences, and uncertainties of the residents and authorities, even though they are ostensibly grappling with the same problem. It shows how consequences bearing the same name may have different meaning and different evaluations in the two contexts. Diagnosing these differences can improve communication between the groups by prompting a common nomenclature and improve respect between the groups by showing the potential legitimacy of their individual perspectives. In some cases, the result will be to reduce conflict, by showing differences to be more apparent than real; in other cases, conflicts will be shown in greater relief.

The analysis may also reveal ways in which decisions at the two levels are coupled. For example, if the authorities adopt a strictly enforced standard, then the residents have few decisions to make. Conversely, if the residents act spontaneously, then the authorities have no need to act. On the other hand, if the residents feel that they cannot afford to act, then they may pressure the authorities to add compensation to its option set.

More generally, where there is conflict or coupling, there will often be pressure to change, which can lead to changes in the respective decision problems. One side may attempt to impose its will on the other, thereby eliminating options from its decision tree (e.g. prohibiting the issuance of standards). One side may offer side payments to the other, thereby changing the set of consequences (e.g. subsidizing ventilation). One side may attempt to persuade the other to change its beliefs, thereby changing the probability and consequence evaluations (e.g. a public education campaign regarding the risks of radon).

The intent of such actions is to render the actions of the overall system more predictable, by getting the other side to take one particular action. In the short run, though, they serve to render analyses like the present one somewhat indeterminate, in so far as the decisions that people eventually face may be different than the ones described. By highlighting the sources of conflict and the possibilities for resolution, the analysis should prompt people to change the facts of the situation. In this sense, such analyses sow the seeds of their own obsolescence or, rather, the need for iterative analyses, each incorporating the changes prompted by its predecessors. If one side is unhappy with how the other side has altered its situation, then it may exert reciprocal pressures, leading to the need for additional iterations.

A final way in which such parallel analyses can foster change is by showing how sensitive the decisions are to different kinds of information (Raiffa, 1968). For residents, the key information seems to be technical (e.g. how great are the costs? how big is the danger?). For the authorities, it seems to be behavioral (e.g. just how will the public respond to particular standards?). For all parties, the analyses show the importance of information that must come from within themselves, namely how to evaluate the relative importance of the different consequences. If the information is created to meet these needs, then there will be a reduction in the most significant uncertainties surrounding these decisions. Thus, although the decision problems will be changed further, it should be in the direction of a more stable representation.

Extensions

Decision trees can be used to examine the structure of other kinds of environmental disputes as well. Two 'by-products' of the present analyses were descriptions of the individual problems faced by the two parties. Either of these might have been the focus of attention in its own right, with the description of the others' decision being pursued only in so far as it helped with estimating model parameters. Thus, for example, a study of residents' perceptions of random risk might incorporate a side study of the authorities' decision problem as a way of anticipating how their actions might affect the residents' perceptions.

Another use of decision trees would be modeling the position of parties in direct conflict, such as the European nations producing and absorbing acid rain (Svenson and Fischhoff, 1983). Such an analysis might be undertaken (a) by one party to the dispute, in order to get a clearer feeling for how the problem looks 'from the other side' (which might, in turn, suggest compromises that it might offer); (b) by all parties to a dispute in order to clarify and stabilize their own positions prior to negotiations; or (c) by a neutral party hoping to create a comprehensive description of the overall problem which is sensitive to the concerns of all parties.

Whatever kind of dispute is being modeled, the conclusions will depend upon the substantive details of the environmental problem under consideration. For example, automobile seatbelt usage can be viewed from the perspectives of both auto safety authorities and individual drivers. As with the radon example, the authorities' problem is to protect the public interest while being constrained by the public's willingness to act (or to be manipulated). However, unlike the radon example, conflict arises here because the problem is simply not much of an issue for most of the individuals at risk. To the extent that drivers make their seatbelt decisions on a trip-by-trip basis, the risks that they see themselves facing are negligible, making the benefits provided by seatbelts appear small relative to the inconvenience that they cause. These perceived risks are further minimized by the tendency for most drivers to view themselves as being safer than average (Svenson, 1981). By contrast, safety authorities see the overall casualty toll arising from the aggregation of all those seemingly safe trips by seemingly safe drivers. One way to bridge this gap in perceptions is to get individual drivers to adopt something like the broader perspective of safety officials. Although individuals may never care very much about the overall death toll (caused by all those other, unsafe drivers), they may be induced to see the cumulative risk from all of the seemingly safe trips that they take in their lifetime (Slovic *et al.*, 1978). Here, too, modeling decision problems at different levels can suggest ways to change those problems and possibly improve the resolution of environmental risks.

Notes

(1) These heating expenses could be reduced by installing heat exchangers, which clean the air while retaining heat. If the costs of such systems are paid immediately then the balance between immediate and long-term costs shifts from that in Figure 1. If they are amortized over a period of years, then the picture remains the same.

(2) The 'no standard' option would, in effect, return these decisions to the market-place, where one might expect a tendency to let the rich live and the poor die, in so far as the former are more likely to take voluntary action to reduce risk and might do so by selling their radon-contaminated homes to the latter.

Acknowledgements

Support for this research was provided by the Swedish Council for Research in the Humanities and Social Sciences, Swedish Energy Research and Development Commission and the United States National Science Foundation. We thank Kerstin Meyerhoffer, Rocio Klingsell and Leisha Sanders for help in producing the manuscript.

References

Abelson, R. (1985). Decision making. *Handbook of Social Psychology*, 3rd Edit. Reading, Massachusetts: Adison-Wesley.

Baum, A., Gatchell, R. J. and Schaeffer, M. A. (1983). Emotional, behavioral, and physiological effects of chronic stress at Three Mile Island. *Journal of Consulting and Clinical Psychology*, **51**, 565–572.

Chen, K. (1980). *Value-oriented Societal Decision Analysis*. Ann Arbor, Michigan: The University of Michigan.

Cotgrove, S. (1982). *Catastrophe or Cornucopia: The Environment, Politics, and the Future*. New York: Wiley.

Derr, P., Goble, R., Kasperson, R. E. and Kates, R. W. (1983). Responding to the double standard of worker/public protection. *Environment*, **25**, 6–11, 36–36.

Edwards, W. (1980). Reflections on and criticism of a highly political multiattribute utility analysis. In L. Cobb and R. M. Thrall (eds), *Mathematical Frontiers of Behavioral and Policy sciences*. Boulder, Colorado: Westview.

Elliot, G. R. and Eisdorfer, C. (1982). *Stress and Human Health*. New York: Springer.

Fischhoff, B. (1984). Setting standards: A systematic approach to managing public health and safety risks. *Management Science*, **30**, 823–843.

Fischhoff, B. (1985). Judgmental aspects of risk analysis. *Office of Management and Budget Handbook of Risk Analysis*. Washington, D.C.: National Science Foundation.

Fischhoff, B. and Beyth-Marom, R. (1983). Hypothesis evaluation from a Bayesian perspective. *Psychological Review*, **83**, 237–265.

Fischhoff, B., Slovic, P. and Lichtenstein, S. (1981). Lay foibles and expert fables in judgments about risk. In T. O'Riordan and R. Turner (eds), *Progress in Environmental Management*, Vol. 3. London: Academic Press.

Fischhoff, B., Slovic, P. and Lichtenstein, S. (1983). The 'public' versus the 'experts': Actual versus perceived disagreements about the risks of nuclear power. In V. Covello, G. Flamm, J. Rodericks and R. Tardiff (eds), *Analysis of Actual vs. Perceived Risk*. New York: Plenum.

Fischhoff, B., Svenson, O. and Slovic, P. (1985). Active response to environmental hazards: Perceptions and decision making. In D. Stokols and I. Altman (eds), *Handbook of Environmental Psychology*. New York: Wiley (in press).

Fishburn, P. (1982) *Foundations of Expected Utility Theory*. Dordrecht, The Netherlands: D. Reidel.

Gardiner, P. and Edwards, W. (1975). Public values: Multiattribute utility measurement for social decision making. In M. F. Kaplan and S. Schwartz (eds), *Human Judgment and Decision Processes*. New York: Academic Press.

Green, C. H. (1982). Risk: Attitudes and beliefs. In D. V. Canter (ed.), *Behaviour in Fires*. Chichester: Wiley.

Gricar, B. G. and Baratta, A. J. (1983). Bridging the information gap at Three Mile Island: Radiation monitoring by citizens. *Journal of Applied Behavioral Science*, **19**, 35–49.

Hammond, K. R. and Adelman, L. (1976). Science, values and human judgment. *Science*, **194**, 389–396.

Kemp, R., O'Riordan, T. and Purdue, M. (1984). Investigation as legitimacy: The maturing of the big public inquiry. *Geoforum*, **15**, 477–488.

Kunreuther, H., Ginsberg, R., Miller, L., Sagi, P., Slovic, P., Borkan, B. and Katz, N. (1978). *Disaster Insurance Protection: Public Policy Lessons*. New York: Wiley.

Lichtenstein, S., Slovic, P., Fischhoff, B., Layman, M. and Combs, B. (1978). Judged frequency of lethal events. *Journal of Experimental Psychology: Human Learning and Memory*, **4**, 551–578.

Marcus, A. A., Nadel, M. V. and Merrikin, K. (1984). The applicability of regulatory negotiation to disputes involving the Nuclear Regulatory Commission. *Administrative Law Review*, **36**, 213–238.

Murphy, A. H. and Winkler, R. C. (1985). Probability of precipitation forecasts. *Journal of the American Statistical Society*, in press.

Raiffa, H. (1968). *Decision Analysis*. Reading, Massachusetts: Addison–Wesley.

Sjöberg, L. and Jansson, B. (1982). *Boendes uppfattningar om och reaktioner på eventuell förekomst av strålning från radongas in bostaden*. Department of Psychology, University of Göteborg.

Slovic, P., Fischhoff, B. and Lichtenstein, S. (1978). Accident probabilities and seatbelt usage. *Accident Analysis and Prevention*, **10**, 281–285.

Slovic, P., Fischhoff, B. and Lichtenstein, S. (1979). Rating the risks. *Environment*, **21**, 14–20, 36–39.

Slovic, P., Lichtenstein, S. and Fischhoff, B. (1985). Decision making. In R. C. Atkinson, R. Herrnstein, D. Krantz and P. Suppes (eds), *Stevens' Handbook of Experimental Psychology*, 2nd Edit. New York: Wiley (in press).

Statens Strålskyddsinstitut (1982). *Reviderade synpunter på radongränser inför radonutredningens slutrapport*. Report Number 03/149/82. Stockholm: National Institute of Radiation Protection.

Svenson, O. (1981). Are we all less risky and more skillful than our fellow drivers? *Acta Psychologica*, **47**, 143–148.

Svenson, O. and Fischhoff, B. (1983). *Acid rain: case study of a decision problem*. Report to the Swedish Energy Research and Development Commission. Department of Psychology, University of Stockholm.

Ungson, G. and Braunstein, M. (1982). *Decision making: An interdisciplinary approach*. New York: Kent.

Vlek, C. and Stallen, P.-J. (1980). Rational and personal aspects of risk. *Acta Psychologica*, **45**, 273–300.

Weinstein, N. D. (1980). Unrealistic optimism about future life events. *Journal of Personality and Social Psychology*, **39**, 806–820.

PUBLIC ATTITUDES TO NUCLEAR ENERGY: SALIENCE AND ANXIETY

JOOP VAN DER PLIGT*

University of Exeter, Exeter, U.K.

Abstract

During the last decade there has been a significant increase in public concern about nuclear energy. This paper presents a brief overview of trends and developments in public opinion since the late 1970s. One possible reason for this increased concern is the public's perception of risks. Research has shown a considerable divergence in public and expert assessment of the risks associated with nuclear energy. It will be argued that qualitative aspects of these risks play a crucial role in the public's perception of nuclear energy, and that reactions such as fear and anxiety are the major determinants of attitudes to the building of new nuclear power stations in one's neighbourhood. It is also clear, however, that differences in the perception of these risks do not embrace all the relevant aspects of public acceptance of nuclear energy. Public reaction is also related to more general values and beliefs, and the issue of nuclear energy is firmly embedded in a much wider moral and political domain.

Introduction

The economic and political significance of energy supplies has dramatically increased since the 1973–74 energy crisis. Until then the debate on nuclear energy had largely centred on 'technical arguments about technical issues' (White, 1977, p. 647). Changes in public awareness and involvement have led to the recognition that public acceptability of nuclear energy will play an important role in the future of this technology.

Over the past decade public support for nuclear energy has gradually been eroded. This increased opposition to nuclear energy was accompanied by the growth of the environmental movement. Since the mid-1970s the environmental movement has matured in organization, has broadened its membership and gained in political significance. Media interest in environmental issues has also increased. The press has shown substantial concern over nuclear safety; a number of accidents and the issue of nuclear waste have been widely reported. Although there is no necessary causal link between media coverage and public attitudes, the increase in media attention was accompanied by increasing public concern over potentially catastrophic accidents and radioactive wastes.

Various developments reflect this increase in public concern. For instance, several national referenda on the issue of nuclear energy have been decided by very narrow margins (Austria, Switzerland and Sweden). Other countries organized national discussion and/or information campaigns (e.g. Austria, 1976–77; The Netherlands, 1982–83) in an attempt to involve further the public in the nuclear debate. The increasing length of public inquiries into the issue of nuclear energy is another example

* The author is now at Free University, Institute for Environmental Studies P.O. Box 7161, 1007 MC Amsterdam, The Netherlands.

of the public's concern about this technology. Finally, public concern is clearly reflected in the growing number of public opinion polls on the issue. The next section presents a brief overview of developments in public opinion in both the U.S.A. and Europe.

Public Opinion

Prior to the mid-1970s survey data showed consistently high levels of support for nuclear energy. The place of nuclear energy as a source of electrical power seemed assured. This majority was eroded in the 1970s. Although the slippage was apparent prior to the Three Mile Island accident in 1979, it has been accelerated by that event. Immediately following the Three Mile Island accident, support in the U.S.A. decreased, uncertainty about taking a stand on nuclear power decreased and opposition towards nuclear power increased. Although there has been some rebound towards pre-Three Mile Island levels of support and opposition, the return has not been complete. Recent figures show that the percentage of the U.S.A. public that supports the continued building of nuclear power plants in the U.S.A. is, on average, 5–10% more than the percentage of the public that opposes such construction. This small pro majority is composed of a strong *majority* of men and a *minority* of women. Furthermore, a majority of the public believes that more such accidents are likely to happen (Rankin *et al.*, 1981). Finally, a large majority of the public (about 80%) now says that it is concerned about waste management issues (Kasperson *et al.*, 1980).

The above trends are also apparent in Europe and the U.K. Public opinion in The Netherlands has shown an 'anti-nuclear' majority since the late 1970s. Opinion poll data for the U.K. show a slow but steady increase in public opposition to nuclear energy since the mid-1970s. Whereas in 1980 there was hardly any difference between the number of opponents and supporters of nuclear energy, a National Opinion Poll survey conducted in October 1981 showed that 33% of the public was in favour of expanding the number of nuclear power stations in the U.K., while 53% were opposed.

Recent EEC surveys provide a more complete picture. Table 1 gives an overview of public opinion trends in the 10 member states since 1978. Averaged over the 10 member states 38% of the public favours expansion of nuclear energy, with 37% opposing further development. In 1978 these figures were 44 and 36%, respectively.

Results of public opinion surveys in both the U.S. and Europe show that people are less willing to approve construction of a new reactor close to their community than to approve the construction of these energy facilities in general. Support for local nuclear power plants has been in decline since the mid-1970s. In the U.S.A. support decreased from 47% in 1977 to 28% in 1980 (Rankin *et al.*, 1981). Our survey of three small communities in the South West of England confronted with the possible building of a nuclear power station showed a considerable majority opposing the plans (van der Pligt *et al.*, 1985).

A number of surveys have either compared levels of acceptance of a nuclear power plant amongst people who live near one with that of people who do not, or they have monitored the local climate of opinion in a locality where the possibility of a nuclear power plant being constructed gradually becomes a reality. Overall, there is mixed support for the idea that familiarity leads to greater acceptance of a nuclear

TABLE 1

Attitudes to further development of nuclear power in the 10 EEC member states[a]

| | Percentage of respondents indicating worth of further development | | | | | |
| | Worthwhile | | | Unacceptable risk | | |
	1978	1982	Difference	1978	1982	Difference
Belgium	29	27	− 2	39	37	− 2
Denmark	37	25	− 12	34	49	+ 15
France	40	51	+ 11	42	31	− 11
Federal Republic of Germany	35	37	+ 2	45	30	− 15
Greece	—	15	—	—	50	—
Ireland	43	13	− 32	35	47	+ 12
Italy	53	34	− 19	29	43	+ 14
Luxembourg	35	32	− 3	31	49	+ 18
The Netherlands	28	34	+ 6	54	48	− 6
United Kingdom	57	39	− 18	25	37	+ 12

[a] Total sample size 9700 (1982). Greece was not included in the 1978 survey. Respondents were presented with four response categories to indicate their attitude concerning the development of nuclear power stations, these were: 'it is worthwhile', 'no particular interest', 'the risks involved are unacceptable' and 'don't know'. This table is based on findings reported by the Commission of the European Communities (1982).

power plant in one's community. Melber *et al.* (1977) mentions eight studies which followed local acceptance of a nuclear power plant as it was being constructed. Only two found a significant increase in acceptance over time, and one locality showed a significant increase in the level of opposition. Results are equally mixed concerning the relationship between living near a nuclear power plant and acceptance of nuclear energy in general (see Thomas and Baillie, 1982). Our own surveys (Reference Note 1) showed a marginally more favourable attitude towards nuclear energy in general around Hinkley Point (the site of two existing nuclear power stations in the South West of England) than in three small local communities that were shortlisted by the CEGB as possible future sites. Other research on local attitudes did not support the notion that familiarity leads to more favourable attitudes (e.g. Warren, 1981).

Unfortunately, opinion poll data are usually based on one simple question on the issue of nuclear energy and do not allow clear and firm conclusions about the various beliefs underlying public attitudes. The influence of the wording of opinion polls provides a further cautionary note (see, for example, Roiser, 1983). The opinion polls do show, however, that the public is divided, both in Europe and the U.S.A. Research on public perception of the various benefits and risks of nuclear energy is likely to provide further information on public concern about this technology. A number of surveys indicated that people's perception of risks is an important component of the public's concern over nuclear energy. The issue of risk perception has been extensively studied in recent years, partly with the aim of helping to formulate policy decisions on risk regulation and risk-bearing technologies.

Risk Perception

Although the experts' assessment of the risks of nuclear energy indicate that these are no greater than, and perhaps substantially less than, those of other generally accepted technologies, the public distrust of nuclear energy is substantial. Opinion polls consistently report qualms about the release of radioactivity, potential catastrophic accidents and the disposal of nuclear waste. Both operational hazards and possible adverse environmental impact are seen as major risks of nuclear energy. Our own research in the South West revealed a similar pattern; the nuclear waste issue and risks to the environment played a major role in public perception and acceptability of the building of a nuclear power station.

It is clear that the lay public defines risks in much broader terms than the expert. Early research on risk perception aimed to discover the basis of the public's distrust, given expert assessments of the extreme low probability of serious accidents and the negligible consequences of routine emission to both health and the environment. The experts' risk assessments were regarded as objective and quantifiable, and public fears were interpreted as biased and irrational. Public disagreement among scientists over the risks of nuclear energy, however, led to the realization that even the experts' assessments are less 'objective' than previously assumed. Recent research has paid more attention to the study of how people think about risks. A number of studies have revealed that nuclear power, as compared with other technologies, elicits an extraordinary level of concern, particularly because of the characteristics of the hazards that it poses (see, for example, Fischhoff et al., 1978, 1981). Most prominent among these are the potentially catastrophic and involuntary nature of possible accidents, and the fact that it is an unknown hazard. Compared to other technologies nuclear energy emerges as the most extreme in terms of the size and seriousness of a potential accident.

The public's concept of risk, therefore, seems to be heavily influenced by the characteristics and seriousness of the possible consequences of nuclear energy. These factors play a more important role than the assumed probability of the possible negative consequences.

The concept of risk, however, does not embrace all the relevant terms of public acceptance. The public's perceptions of risks are built on values, attitudes and sets of attributes which need not be similar to the representations of the experts and policy makers.

Beliefs and Values

Attempts to analyse the structure of people's attitudes towards nuclear energy are usually based on expectancy-value models of attitude formation, which broadly assume that the more a person believes the attitude object has good rather than bad attributes or consequences, the more favourable his or her attitude tends to be. In other words, people's attitudes towards nuclear energy are assumed to be a function of beliefs about the possible consequences of its use. Most of the work in this area is based on the expectancy-value model of attitude formation proposed by Fishbein and his colleagues (Fishbein, 1963; Fishbein and Hunter, 1964), which analyses attitudes in relation to the anticipated consequences accompanying the attitude object. Results of these studies show that individual attitudes are based upon perception of various potential negative and positive aspects of nuclear energy (e.g. Otway and Fishbein, 1976; Sundstrom et al., 1977, 1981).

A further conclusion of this research is that separate dimensions of the issue of nuclear energy appear differentially salient for different attitude groups. Otway *et al.*, (1978) report the results of a factor analysis on 39 belief statements about nuclear energy. Results of this work pointed at a number of dimensions underlying the way people think about nuclear energy. Otway *et al.*, (1978), summarized these dimensions as follows: (1) beliefs about the economic benefits of nuclear power, (2) beliefs about environmental and physical hazards due to routine low-level radiation, and possible accidents, (3) beliefs about the socio-political implications of nuclear power (e.g. restrictions on civil liberties), and (4) beliefs about psychological risks (fear, stress, etc.). Subgroups of the 50 most pro- and 50 most anti-nuclear respondents were then compared in order to determine the contribution of each of the four factors to respondents' overall attitudes. For the pro-nuclear group, the economic and technical benefits factor made the most important contribution, whereas for the anti-nuclear group, the risk factors were more important.

Woo and Castore (1980) also found that nuclear proponents attached greater value to the potential benefits of nuclear energy, while the nuclear opponents were more concerned with potential health and safety issues. Results obtained by Eiser and van der Pligt (1979), and van der Pligt *et al.*, (1982), provide further support for the view that individuals with opposing attitudes tend to see different aspects of nuclear energy as salient and, hence, will disagree not only over the likelihood of the various consequences but also over their importance. In other words, each group has its own reasons for holding a particular attitude; the proponents stressing the importance of economic benefits, while the opponents attach greater value to environmental and public health aspects. An important finding of these studies was that the overall attitude of respondents was more closely related to ratings of—in their view—important aspects than to their ratings of subjectively less important aspects. Thus, a consideration of both the perception of the various consequences and the subjective importance or salience provides a more complete picture than could be obtained from a consideration of either factor alone. These studies (Eiser and van der Pligt, 1979; van der Pligt *et al.*, 1982) suggest that the attitudinal differences apparent in controversies of this kind require a conception of attitudes that takes account of the fact that different aspects of the issue will be salient to the different sides of the debate, and that such differences in salience may be at least as clear-cut and informative as differences in the likelihood and evaluation of the various potential consequences. As argued elsewhere (van der Pligt and Eiser, 1984), the finding that separate dimensions of the issue appear differentially salient (both subjectively and in their contributions to the prediction of overall attitude) for the different attitude groups, has important practical implications for theories of attitude and our understanding of why people hold different attitudes towards nuclear energy.

The above studies, however, focus on public attitudes towards nuclear energy in general. More recently we attempted to investigate the relationships between people's attitudes towards the building of a nuclear power station in their locality, their specific beliefs about the local consequences and their perception of the importance of these consequences. This study was conducted in localities that were short-listed by the Central Electricity Generating Board as possible locations for a new nuclear power station in South West England. The sample of respondents was drawn from the electoral register for three communities which were close to the three possible locations. A total of 450 people received a questionnaire by mail. Of this sample,

24 respondents had moved from the area and 300 persons agreed to participate in the study; a response rate of 70%. The sample contained an equal number of males and females, and most respondents had lived for a considerable time in the area (nearly 30 years, on average).

The questionnaire was closed-ended and was preceded by a short introduction describing the Central Electricity Generating Board's announcement concerning the possible locations for the next nuclear power station in the South West of England. The questionnaire assessed subjects' attitudes towards building more nuclear power stations in the U.K., the South West of England and their locality. Respondents were also asked to indicate their attitude towards various other industrial developments in their locality and, finally, to indicate their attitude towards nuclear energy in general, their involvement with the issue, whether they had attended any public meetings on the issue and which aspects should receive most attention in a public inquiry on the possible building up of a new nuclear power station. A more detailed description of the study can be found in van der Pligt et al., (1985).

In the present context we will summarize the findings concerning the perception of the various potential costs and benefits of a nuclear power station in one's locality. Participants were generally opposed to the construction of a nuclear power station in their neighbourhood. It needs to be added, however, that most respondents were also opposed to other industrial activities such as the building of a chemicals plant. These findings indicate a more negative attitude towards large-scale industrial developments than those obtained in some of the literature on the impacts of rapid growth of communities (e.g. Freudenburg, 1984).

In order to investigate people's perception of the various potential costs and benefits of a nuclear power station we presented subjects with two sets of 15 potential consequences. The first set contained 15 immediate effects of the building and operation of a nuclear power station in the locality, while the second set focused on long term consequences. Subjects were split into three attitude groups on the basis of their answer to the question whether they were opposed to or in favour of the building of a new nuclear power station in their locality. A discriminant analysis revealed that the three attitude groups (pro, neutral and anti) differed significantly in their assessment of a number of immediate consequences. The aspects that were most differentially perceived concerned the area of land fenced off, the conversion of land from agricultural use and the prospect of workers coming into the area. Opponents generally thought these developments to have an adverse impact on the quality of life in the locality, while proponents thought the impact of these factors would be relatively minimal.

We also asked people to select the five (out of 15 immediate consequences) aspects they thought to be the most important. Results showed three aspects that were rated very differently as a function of own attitude. Of the pro subjects, 67% regarded road building an important aspect, while only 20% of the anti subjects selected this item among the five most important. A similar difference was obtained concerning the prospect of workers coming into the area (53% of the pros and 18% of the antis). The antis, on the other hand, attached greater importance to the possible conversion of land from agricultural use than the pros (58% vs. 27%).

The mean ratings by the three attitude groups of the 15 (mainly long-term) effects of the building and operation of a nuclear power station in their neighbourhood also showed highly significant differences. Again, we conducted a discriminant

TABLE 2

Perceived consequences of the building and operation of a nuclear power station as a function of attitude

	Impact[a]			Importance[b]		
	Pro (N = 30)	Neutral (N = 40)	Anti (N = 209)	Pro (N = 30)	Neutral (N = 40)	Anti (N = 217)
Economic factors						
Employment opportunities	8·3	7·6	6.0[c]	73	57	15[c]
Business investment	6·7	6·0	4·2	27	28	11
Environmental factors						
Wildlife	4·7	2·6	1·5	40	57	67
Marine environment	5·1	3·6	2·3	13	17	38
Farming industry	4·6	3·0	1·9	17	45	56
Landscape	4·3	2·6	1·3	23	50	66
Public health and psychological risks						
Health of local inhabitants	5·0	4·4	2·6	20	29	48
Your personal peace of mind	5·4	4·3	1·7	27	17	47
Social factors						
Social life in the neighbourhood	6·9	5·7	3·9	30	7	11
Standard of local transport and social services	6·8	6·3	4·9	40	17	5
Standard of shopping facilities	6·6	5·8	4·9	20	14	4

[a] Possible range of scores from 1 (consequence will affect life in the neighbourhood very much for the worse) to 9 (very much for the better).

[b] The scores represent the percentage of subjects selecting each factor among the five most important.

[c] The differences between the three attitude groups were significant in all cases ($P < 0.05$) as indicated by the linear F-term.

analysis to find out which aspects most distinguished the three attitude groups. The results revealed three aspects which had considerable predictive power in separating the three attitude groups. These were the perceived effects on one's 'peace of mind' and the effects on the environment and wildlife. The first aspect corresponds to what Otway *et al.* (1978) called 'psychological risk', while the other two aspects are related to what these authors termed 'environmental and physical risk'.

Again we asked the respondents to choose the five consequences they regarded most important. The results showed very marked differences between the three attitude groups. The most striking difference concerned the possible effects on employment opportunities, 73% of the pros selected this item among the most important, while only 15% of the antis considered this aspect as important. Overall, the pro respondents stressed the importance of economic benefits, while the antis stressed the risk factors (both environmental and psychological risks). Table 2 presents a summary of these differences in the perception of the various long-term consequences and their importance. A closer inspection of these differences underlines the importance of including both beliefs and salience in one's conception of attitude. Even though the attitude groups, for example, showed relatively minor differences in their evaluation of the effects of potential employment opportunities in the locality, a majority of the pros found this aspect important, while only a small minority of the antis regarded this aspect as being of importance.

Results of this study showed that the major differences between the attitude groups concerns the less tangible, more long-term nature of the potential negative outcomes. Our findings further suggested that the perception of the psychological risks are the prime determinant of attitudes as indicated by the very high correlation (0·80) between this factor and attitude towards the building of a new nuclear power station. Other studies (e.g. Woo and Castore, 1980) did not find such a strong relation between psychological risks and attitude. One reason for this could be that our research concentrated on people living very close to the proposed nuclear power station (all within a five-mile radius). Most other studies used much wider areas around proposed nuclear power stations.

In summary, opponents and proponents of nuclear energy have *very different* views on the possible consequences of nuclear energy. Our research indicates that this applies to both the general issue of nuclear energy and to the building of a nuclear power station in one's locality. The most significant difference, however, concerns the perception of psychological risks (anxiety, stress). This factor becomes more important when people are (or will be) more directly exposed to the risks, for instance when their locality is shortlisted as a possible site for a nuclear power station.

These findings also suggest that the different perceptions of the possible consequences of further expansion of the nuclear industry are related to more general values. Eiser and van der Pligt (1979) addressed this point and asked 47 participants attending a one-day workshop on 'The Great Nuclear Debate' to select the five factors which they felt 'would contribute most to an improvement in the overall "quality of life"' from a list of nine. Table 3 summarizes some of their findings.

Results showed marked differences between the two groups, with pro-nuclear subjects stressing the importance of 'advances in science and technology', 'industrial modernization', 'security of employment' and 'conservation of the natural environment'. The anti-nuclear respondents put even more emphasis on the last factor and stressed the importance of 'decreased emphasis on materialistic values', 'reduction in scale of industrial, commercial and governmental units' and 'improved social welfare'.

TABLE 3

The importance of general values as a function of attitude[a]

	Percentage of respondents selecting each factor	
	Pro-nuclear subjects	Anti-nuclear subjects
Decreased emphasis on materialistic values	36	100
Reduction in scale of industrial, commercial and governmental units	22	86
Industrial modernization	68	6
Security of employment	77	40
Improved social welfare	31	80
Conservation of the natural environment	77	100
Advances in science and technology	82	13

[a] Adapted from Eiser and van der Pligt (1979), p. 532.

More recently, we presented a sample of the Dutch population with a similar list of more general values (van der Pligt *et al.*, 1982). Results were in accordance with the above study and showed that pro-nuclear respondents stressed the importance of economic development, whereas anti-nuclear respondents put more emphasis on conservation of the natural environment and the reduction of energy use. Finally, the anti-nuclear group thought the issue of increased public participation in decision making to be more important.

Not surprisingly, we also found a relation between these value differences and respondents' position on a political left–right dimension. Political preference was significantly related to these differences in values, and to the respondents' attitudes to the building of more nuclear power stations in The Netherlands. Opinion poll surveys conducted in The Netherlands when this study was being carried out confirmed this relationship between political preference and attitudes towards nuclear energy (see van der Pligt *et al.*, 1982).

Discussion

The present research has demonstrated that the public is divided over the issue of nuclear energy. Public opinion data from the U.S.A. and Europe show marginal differences between the percentages of people opposed to further expansion of this technology and those in favour. Survey data also indicated that safety issues play a crucial role in public attitudes. Research on public perception of risks has pointed out a number of characteristics of the risks of nuclear energy that elicit an extraordinary level of concern.

Both public disagreement among scientists concerning the likelihood and magnitude of potential risks of reactor operation and waste storage and the frequent mention of possible health hazards will reinforce public distrust of nuclear technology. Since safety-related issues play a crucial role in public acceptance of this technology, it seems necessary to improve the relations between the expert and the lay public. For the lay public this poses an important challenge: to be better informed and to be aware of the qualitative aspects that strongly affect their perception of risks. For experts it seems necessary to recognize the limitations and fallibility of risk assessments, and to be aware of the fact that important qualitative aspects of risks influence the responses of lay people. However, it is also clear that the risk concept in itself is not sufficient to explain public reactions. Risk perception is not the only issue of importance. Public reaction is also related to more general beliefs and values, and the issue of nuclear energy is firmly embedded in a much wider moral and political domain.

Individuals with opposing attitudes tend to see different aspects of the issue as salient, and hence, will disagree not only over the likelihood of the various consequences but also over their importance. In other words, *opponents and proponents have different reasons for holding their particular attitudes.* The supporters see the potential economic benefits as most important, while the opponents attach greater value to environmental and public health aspects. Present findings further suggest that people's attitude towards the building of a new nuclear power station in their neighbourhood is very closely related to their perception of the psychological risks of such a development.

Results concerning the perception of the importance of more general social issues were in line with the above findings, and indicated that attitudinal differences towards nuclear energy are embedded in a wider context of attitudes towards more general social issues. Public thinking on nuclear power is not simply a matter of perceptions of risks but is also related to more generic issues such as the value of economic growth, high technology and centralization. It seems impossible, therefore to detach the issue of nuclear energy from questions of the kind of society in which one wants to live.

The fact that public attitudes are relatively stable *and* embedded in a wider context of values suggests that large-scale attitude conversion may be more difficult than often assumed. People may, however, change their attitudes as a function of serious accidents that attract widespread attention, especially if they have not committed themselves strongly to one of the two sides. With regard to safety-related aspects of public acceptance of nuclear power, it seems much easier for nuclear attitudes to become suddenly more anti-nuclear because of a major accident or a series of smaller accidents (e.g. the recent events at the Sellafield reprocessing plant, see Reference Note 2) than it would be for nuclear attitudes to become more pronuclear as a longer-term result of an extensive period of safe operations. Changes in a pro-nuclear direction are more likely to result from events related to energy supply, e.g. developments that would make non-nuclear energy much more expensive.

Reference Notes

(1) van der Pligt, J., Eiser, J. R. and Spears, R. (1984). Attitudes toward the building of a nuclear power station: familiarity and salience. Paper submitted for publication.
(2) van der Pligt, J., Eiser, J. R. and Spears, R. (1984). Nuclear energy: accidents and attitudes. Paper submitted for publication.

References

Commission of the European Communities. (1982). *Public opinion in the European Community*. Report No. XVII/202/83-E. Brussels, Belgium: Commission of the European Communities.

Eiser, J. R. and van der Pligt, J. (1979). Beliefs and values in the nuclear debate. *Journal of Applied Social Psychology*, **9**, 524–536.

Fischhoff, B., Lichtenstein, S., Slovic, P., Derby, S. L. and Keeney, R. L. (1981). *Acceptable Risk*. Cambridge: Cambridge University Press.

Fischhoff, B., Slovic, P., Lichtenstein, S., Read, S. and Combs, B. (1978). How safe is safe enough: a psychometric study of attitudes towards technological risks and benefits. *Policy Sciences*, **8**, 127–152.

Fishbein, M. (1963). An investigation of the relationship between beliefs about an object and the attitude towards that object. *Human Relations*, **16**, 233–240.

Fishbein, M. and Hunter, R. (1964). Summation versus balance in attitude organization and change. *Journal of Abnormal and Social Psychology*, **69**, 505–510.

Freudenburg, W. R. (1984). Boomtown's youth: the differential impact of rapid community growth on adolescents and adults. *American Sociological Review*, **49**, 697–705.

Kasperson, R. E., Berk, G., Pijawka, D., Sharaf, A. B. and Wood, J. (1980). Public opposition to nuclear energy: Retrospect and prospect. *Science, Technology and Human Values*, **5**, 11–23.

Melber, B. D., Nealey, S. M., Hammersla, J. and Rankin, W. L. (1977). *Nuclear Power and the public: analysis of collected survey research*. Battelle Memorial Institute, Human Affairs Research Centres, Seattle, Washington.

Otway, H. J. and Fishbein, M. (1976). *The determinants of attitude formation: an application to nuclear power*. Research Memorandum RM-76-80. Laxenburg, Austria: International Institute for Applied Systems Analysis.

Otway, H. J., Maurer, D. and Thomas, K. (1978). Nuclear power: The question of public acceptance. *Futures*, **10**, 109–118.

Rankin, W. L., Melber, B. D., Overcast, T. D. and Nealey, S. M. (1981). *Nuclear power and the public: an update of collected survey research on nuclear power*. PNL-4048. Batelle Human Affairs Research Centres, Seattle, Washington.

Roiser, M. (1983). The uses and abuses of polls: A social psychologist's view. *Bulletin of the British Psychological Society*, **36**, 159–161.

Sundstrom, E., Lounsbury, J. W., DeVault, R. C. and Peelle, E. (1981). Acceptance of a nuclear power plant: applications of the expectancy-value model. In A. Baum and J. E. Singer, *Advances in Environmental Psychology*, Vol. 3. Hillsdale, New Jersey: Lawrence Erlbaum, pp. 171–189.

Sundstrom, E., Lounsbury, J. W., Schuller, C. R., Fowler, J. R. and Mattingly, T. J., Jr. (1977). Community Attitudes towards a proposed nuclear power generating facility as a function of expected outcomes. *Journal of Community Psychology*, **5**, 199–208.

Thomas, K. and Baillie A. (1982). *Public attitudes to the risks, costs and benefits of nuclear power*. Paper presented at a joint SERC/SSRC seminar on research into nuclear power development policies in Britain, June, 1982.

van der Pligt, J. and Eiser, J. R. (1984). Dimensional salience, judgment and attitudes. In J. R. Eiser (ed.), *Attitudinal judgment*. New York: Springer, pp. 161–177.

van der Pligt, J., van der Linden, J. and Ester, P. (1982). Attitudes to nuclear energy: Beliefs, values and false consensus. *Journal of Environmental Psychology*, **2**, 221–231.

van der Pligt, J., Eiser, J. R. and Spears, R. (1985). Construction of a nuclear power station in one's locality: Attitudes and salience. *Basic and Applied Social Psychology*, in press.

Warren, D. S. (1981). *Local attitudes to the proposed Sizewell 'B' nuclear reactor*. Report RE 19. Food and Energy Research Centre, October, 1981.

White, D. (1977). Nuclear Power: A special new society survey. *New Society*, **39**, 641–650.

Woo, T. O. and Castore, C. H. (1980). Expectancy value and selective exposure as determinants of attitudes toward a nuclear power plant. *Journal of Applied Social Psychology*, **10**, 224–234.

NUCLEAR ATTITUDES AFTER CHERNOBYL: A CROSS-NATIONAL STUDY

J. RICHARD EISER,* BETTINA HANNOVER,† LEON MANN,‡
MICHEL MORIN,§ JOOP VAN DER PLIGT,‖ and PAUL WEBLEY*

*Department of Psychology, University of Exeter, Exeter EX4 4QG, U.K.;
†Technical University, Berlin; ‡Flinders University of South Australia;
§University of Provence; and ‖University of Amsterdam

Abstract

A total of 840 subjects from universities in Australia, England, France, Germany and The Netherlands completed a questionnaire during the months following the Chernobyl accident. Items included measures of political decision-making style, nuclear attitudes, reactions to Chernobyl and general political orientation. Decision-making style and the favourability/unfavourability of nuclear attitudes were relatively independent of each other. However, those who described themselves as more informed and interested in nuclear issues, and as having paid more attention to, and having been more frightened by, the news of Chernobyl, scored lower on the style of 'defensive avoidance' but higher on that of 'self-esteem/vigilance'. Reactions to Chernobyl were strongly related to attitudes on other nuclear issues defined within specific national contexts, and more conservative or right-wing political preferences were predictive of greater support for nuclear power.

Introduction

A long tradition of research has been concerned with how people interpret information that is threatening and/or inconsistent with their existing attitudes and beliefs (Festinger et al., 1956; Abelson, 1959). On the one hand, people may resist changing their attitudes even in the face of inconsistent information (Crocker, Fiske & Taylor, 1984). On the other, just having to make ideological and political decisions because of new information or events can be experienced as stressful and give rise to feelings of conflict. According to Janis and Mann (1977), while some individuals may show 'unconflicted adherence' to their existing beliefs and simply ignore inconsistent information, others may adopt a number of cognitive strategies to deal with decision conflicts. Especially important is the strategy of 'defensive avoidance' (not thinking about the problem), which is expressed variously in the forms of 'buckpassing' (letting others decide for one), 'procrastination' (reluctance to make any commitment) and 'rationalization' (distortion and denial of the issues involved in the choice). Other patterns include 'hypervigilance' (panicky or impulsive action on the basis of partial information) and, possibly the most adaptive, 'vigilance' (a weighing-up of the benefits and costs of different options). Janis and Rausch (1970) found a vigilant openness to dissonant information even among draft resisters who had signed a 'We won't go' pledge during the Vietnam war. Other research (Radford et al., 1986; Burnett et al., 1988) relates individuals' evaluations of their own competence as decision-makers to their use of the different styles or strategies for dealing with conflict (particularly 'vigilance').

Conflict theory was initially based largely on data from people's reactions to fear-arousing messages and emerging warnings (Janis, 1951, 1962). The nuclear accident at Chernobyl in April 1986 provided an opportunity to examine the contemporary relevance of the conflict theory analysis of coping patterns to an understanding of reactions to an event of great international importance. The news of the accident was both threatening and uncertain in its implications, both immediately and for a long time afterwards. The Chernobyl accident could be regarded as giving rise to a form of 'dissonance' or 'conflict' in at least two major respects: first, it was inconsistent with more favourable estimations of the safety of nuclear power stations held by many (though obviously not all) ordinary people; second, it raised the question of whether ordinary people could in fact do anything to reduce their own vulnerability to the effects, not only of the Chernobyl accident, but also of possible future nuclear accidents, both far and near. Accordingly, a study of reactions to Chernobyl enables us to examine whether people responded complacently to the threat, or resolved their conflict about what stand or action to take by using one or more of the coping patterns identified by Janis and Mann (1977). The present paper does not deal with the impact of time on attitudes toward the Chernobyl accident, which is studied in several other contributions to this volume; our research emphasizes structural aspects of public reactions and attitudes.

This study considers how young people in different countries (varying considerably in proximity to Chernobyl) reacted to the news of Chernobyl as a function principally of the following factors: their preferred style of coping with decision conflicts in a political context; their attitudes toward nuclear issues and policies in their own country; their views on the functioning of their national economies and their general political orientation. Our general hypotheses were that the implications of the Chernobyl accident would be interpreted in a manner consistent with more general attitudes toward nuclear power, and that these in turn would relate to broader political and economic beliefs. We anticipated that more conservative political preferences and greater optimism about national economic performance (Webley & Spears, 1986; Webley *et al.*, 1988) would be associated with more favourable attitudes to nuclear power in both its civil and military aspects. Following previous research (e.g. Nealey *et al.*, 1983) we also expected more pronuclear attitudes among men than women.

Individual differences in decision-making style were expected to relate to how personally threatening or important the accident was seen to be. Such perceived threat, moreover, could be expected to depend in part on geographical proximity. Previous research has related nuclear attitudes to proximity of residence to routinely functioning and/or proposed nuclear plants (van der Pligt *et al.*, 1986*a*; Eiser *et al.*, 1988) as well as to the Three Mile Island plant after the accident there (Baum *et al.*, 1983) However, we are dealing here with far greater distances than those compared in these previous studies.

Method

Subjects
The sample consisted of 840 individuals (385 males and 455 females: mean age: 23·6 years), recruited from the following cities: Adelaide, Australia ($N = 291$; 109 males, 182

females; mean age: 22·9 years), Aix-en-Provence, France ($N = 151$; 50 males, 101 females; mean age: 23·8 years); Amsterdam, The Netherlands ($N = 100$; 64 males, 36 females; mean age 21·9 years), West Berlin, Germany ($N = 143$; 82 males, 61 females; mean age: 29·0 years) and Exeter, England ($N = 155$; 80 males, 75 females; mean age: 20·5 years). Subjects were mostly university students in psychology or social sciences, recruited on a voluntary basis during lectures or other teaching periods.

There were uncontrolled differences between the groups in how long after the (April 1986) Chernobyl accident they were tested. The data were: Adelaide—July 1986; Amsterdam—February 1987; Aix—March 1987; Berlin—January 1987; Exeter— October 1986.

Questionnaire

In addition to asking subjects to state their sex and age, the questionnaire included the following sections administered (in translation) to all the national groups:

(1) Political decision-making scale: This contained 12 items adapted from the Flinders Decision Making Questionnaire (Mann, 1982), a measure of individual differences in general decision-making style. The modified items were worded to specify *political* decision-making as the particular focus. The text of these items appears in Table 1. Responses were in terms of the categories 'true for me' (scored as 2), 'sometimes true for me' (1) and 'not true for me' (0).

(2) Economic expectations: Subjects rated, on a nine-point scale from 'very pessimistic' (1) to 'very optimistic' (9) how they generally felt about their country's economic future (a) over the next year and (b) over the next 10 years.

(3) Reactions to Chernobyl: Two questions were answered on a scale from 'very little' (1) to 'very much' (9): 'How much attention did you pay to the news about the accident at Chernobyl?' and 'How frightened were you by this news?' Subjects then rated (on a scale from 1 = 'extremely safe' to 9 = 'extremely dangerous') how dangerous or safe they thought it would be to live at the following distances away from a nuclear power station: 1, 2, 4, 8, 16, 32, 64, 128, 256 and 512 km.

Four statements used in a previous study (Eiser, Spears & Webley, 1989) were responded to on a seven-point scale from 'very strongly disagree' to 'very strongly agree'. These were: 'What happened at Chernobyl could easily happen at *any* nuclear power station'; 'I feel sure that there are going to be many more nuclear disasters before very long'; 'It is extremely unlikely that there will ever be another accident as serious as that at Chernobyl'; and 'Compared with the dangers of other kinds of pollution, people's fears of the effects of radiation are out of all proportion'. These were scored so that higher scores represented more 'pronuclear' positions, i.e. 7 to 1 for the first two statements, 1 to 7 for the last two.

(4) Self-ratings: Subjects rated themselves on the following nine-point scales: 'extremely anti-nuclear' (1) to 'extremely pro-nuclear' (9); 'extremely ill-informed about nuclear issues' (1) to 'extremely well-informed about nuclear issues' (9); 'not at all interested in nuclear issues' (1) to 'extremely interested in nuclear issues' (9).

Remaining items in the questionnaire varied for the different groups. These included (a) questions interpretable as providing direct or indirect indications of political preference and (b) a section measuring approval/disapproval (on a seven-point scale from 'very strongly opposed' to 'very strongly in favour') of various nuclear-related policies of specific national relevance. Typically, these items dealt with the construction of nuclear power stations, regionally and/or nationally (except in the Adelaide sample, who, since there are no nuclear power stations in Australia, were asked instead about the resumption of uranium mining in South Australia), nuclear deterrence and the presence of military nuclear bases (cruise missiles). (Note that these military aspects depend on the national context, with the question of independent nuclear deterrence being pertinent only for France and the U.K.) The European samples also were asked about the waste reprocessing plants in their respective countries, and (apart from Berlin) about underground nuclear waste disposal. The Berlin sample was also asked about civil and military nuclear power in East Germany.

Results

Political decision-making style

The twelve items in this section were submitted to a principal components analysis with varimax rotation. This yielded four factors, with eigenvalues of 3·26, 1·42, 1·12 and 1·02, accounting respectively for 27·2, 11·8, 9·3 and 8·5% of the variance. Table 1 shows the factor loadings and the text of the items.

Factor I seems mainly to reflect what Janis and Mann (1977) term 'defensive avoidance', which consists of tendencies to procrastinate (item k), to 'buckpass' (item d) and to rationalize (item e), as well as attempts to escape the difficulties of political

TABLE 1

Principal components analysis, with varimax rotation, of items measuring decision-making style (loadings × 100). Factor I:= 'Defensive avoidance'; Factor II: 'Self-esteem/vigilance'; Factor III:= 'Passivity'; Factor IV:= 'Hypervigilance'

Item	Factor			
	I	II	III	IV
a I avoid thinking about political issues	71	−19	14	−13
b I get very worked up when I have to make a political choice	−12	−11	03	82
c I find politics boring	74	−05	−02	−05
d I prefer to leave political decisions to other people	67	−15	13	−25
e There's nothing I can do that makes any difference politically	63	−06	−27	15
f I feel so discouraged that I give up trying to decide what stand I should take on political issues	55	−01	−04	28
g I feel confident about my ability to make decisions on political issues	−26	79	−11	−17
h I think that I am a good judge of political issues	−25	77	−19	−12
i I like to consider all of the alternatives before making a political choice	14	60	20	41
j It is easy for other people to convince me that their political opinion rather than mine is the correct one	18	−04	62	−06
k I put off taking a stand on political issues	63	−16	31	−13
l I feel I should be more involved politically	−14	−07	72	12

choice (items *a, f*). Factor II seems to be a '*self-esteem*' factor (e.g. item *g*), together with an element of what Janis and Mann term '*vigilance*' (item *i*). Factor III is more difficult to label, but appears to relate to a kind of '*passivity*' (item *j*). Factor IV loads most heavily on item *b*, assumed to reflect what Janis and Mann term '*hypervigilance*'. The fact that item *i* also loads quite highly on this factor suggests that a consideration of *all* political alternatives may involve a hypervigilant (i.e. obsessive-compulsive) quality as well as a careful-thoughtful ('vigilant') quality.

Factor scores on each of these factors were then computed for each individual. A MANOVA was then performed on these scores to look for difference as a function of sex and nationality. The mean scores are shown in Table 2. With the four factor-scores ('Style') treated as a 'within-subjects' variable, this analysis yielded a non-significant effect for Sex ($F(1,815) = 2.61$), a significant effect for Nation ($F(4,815) = 10.33$, $p < 0.001$) and a non-significant Sex × Nation interaction ($F(4,815) = 2.01$). The multivariate effect for Style ($F(3,813 = 1.67$) and Sex × Nation × Style ($F(12,2445) = 1.14$) were non-significant. However, the Sex × Style ($F(3,813) = 7.15$, $p < 0.001$) and Nation × Style ($F(12,2445) = 19.03$, $p < 0.001$) interactions were significant.

Univariate *F*-tests of the Sex effect gave the following values (with $df = 1,815$) for the four 'styles' (factors) respectively: 2·49, *ns*; 8·78, $p < 0.01$; 10·47, $p < 0.001$; 2·33, *ns*. For the Nation effect, the corresponding values (with $df = 4,815$) were: 19·00, $p < 0.001$; 4·16, $p < 0.01$; 7·19, $p < 0.001$; 39·64, $p < 0.001$. As may be seen in Table 2, females were lower on 'self-esteem/vigilance' than males but higher on 'passivity'. Exeter was highest, and Berlin lowest, on 'defensive avoidance'; Amsterdam was highest, and Aix lowest, on 'self-esteem/vigilance'; Exeter was highest, and Aix lowest, on 'passivity'; and Aix was highest on 'hypervigilance', with Berlin and Adelaide particularly low.

Attitude measures
A principal components analysis with varimax rotation was conducted on the remaining attitude measures common to all groups. For this purpose, the four belief statements concerning the consequences of Chernobyl were summed to give a single score, representing a tendency to regard such consequences as less serious. For the total sample, this summated score (hereafter termed '*downplay*') had a Cronbach's alpha of 0·79. Additionally, two derived scores form the danger ratings of different distances were entered into the analysis: 'danger-range' (the dangerousness of 1 km minus that of

TABLE 2
Group means of decision-making style factor scores

	Factor			
	I 'Defensive avoidance'	II 'Self-esteem/ vigilance'	III 'Passivity'	IV 'Hyper- vigilance'
Males	−0·12	0·10	−0·17	−0·02
Females	−0·01	−0·12	0·07	0·09
Adelaide	0·19	0·09	0·11	−0·23
Aix	−0·07	−0·30	−0·31	0·88
Amsterdam	−0·27	0·15	−0·11	−0·05
Berlin	−0·51	0·04	−0·16	−0·29
Exeter	0·34	−0·02	0·21	−0·12

512 km) and 'mean danger' (the mean of all 10 distances). The analysis, summarized in Table 3, yielded three factors, with eigenvalues of 3·23, 1·63 and 1·26, accounting for 32·3%, 16·3% and 12·6% of the variance. Factor 1 appears to be a *'pronuclear'* factor, associated with less fear concerning Chernobyl and nuclear safety generally; factor 2 appears to reflect *'involvement'* (greater attention to Chernobyl, more interest in, and self-rated informedness concerning nuclear power); and factor 3 reflects *'economic optimism'*.

Although the 'economic optimism' items formed a distinct factor, some of the lower loadings suggest an association between economic optimism and pronuclear attitudes. This was borne out by simple correlations. Self-rated attitude correlated 0·12 ($p < 0.001$) with optimism for the next year, and 0·26 ($p < 0.001$) with optimism for the next 10 years. The correlations between these two optimism measures and other variables were, respectively: attention to Chernobyl, -0.04, *ns*, -0.08, $p < 0.05$; frightened by Chernobyl, -0.07, $p < 0.06$, -0.11, $p < 0.01$; 'downplay', 0·11, $p < 0.01$, 0·22, $p < 0.001$; 'danger range', 0·04, *ns*, 0·13, $p < 0.001$; 'mean danger', -0.13, $p < 0.001$, -0.18, $p < 0.001$; informed (about nuclear issues), 0·10, $p < 0.01$, 0·07, $p < 0.06$; interested, -0.08, $p < 0.05$, -0.08, $p < 0.05$.

A MANOVA was then performed to look for group differences in the factor scores derived from this analysis. The multivariate Fs for Sex ($F(3,780) = 26.53$), Nation ($F(12,2346) = 21.31$) and the Sex × Nation interaction ($F(12,2346) = 3.10$) were all significant at $p < 0.001$. The group means are shown in Table 4. Males were more 'pronuclear' (univariate $F(1,782) = 71.16$, $p < 0.001$) and 'involved' ($F(1,782) = 5.60$, $p < 0.02$) but not significantly more 'optimistic' ($F(1,782) = 2.55$, *ns*). Exeter and Aix were most 'pronuclear' and Berlin most 'antinuclear' ($F((4,782) = 28.02$, $p < 0.001$). Amsterdam and Berlin were most 'involved' and Adelaide least so ($F(4·782) = 9.71$, $p < 0.001$). The Amsterdam group was also the most 'optimistic' with the Aix group the least so ($F(4,782) = 28.41$, $p < 0.001$). Univariate tests showed that the significant Sex × Nation interaction arose from differences on the 'pronuclear' factor only ($F(4,782) = 7.43, p < 0.001$). The overall tendency for males to be more 'pronuclear' than females did not occur within the Exeter group (means: males, 0·357, females, 0·415), and was rather weaker within the Berlin group (-0.423, -0.855) than in the remainder (Adelaide, 0·469, -0.361; Aix, 0·810, -0.123; Amsterdam, 0·412, -0.351).

TABLE 3

Principal components analysis, with varimax rotation, of the attitude measures (Loadings × 100)

	Factor		
	I 'Pronuclear'	II 'Involvement'	III 'Economic optimism'
Economic optimism: 1 year	03	00	88
Economic optimism: 10 years	17	−01	84
Attention to Chernobyl	−17	76	−08
Frightened by Chernobyl	−54	41	−07
'Downplay'	82	−09	11
Self-rated attitude	79	−09	16
Informed	04	69	19
Interested	−19	76	−11

TABLE 4
Group means of attitude factor scores

	Factor		
	I 'Pronuclear'	II 'Involvement'	III 'Economic optimism'
Males	0·33	0·16	0·12
Females	−0·25	−0·02	0·01
Adelaide	0·05	−0·21	−0·09
Aix	0·34	−0·06	−0·39
Amsterdam	0·03	0·34	0·88
Berlin	−0·64	0·33	0·06
Exeter	0·39	−0·05	−0·15

Relationships between attitude and decision-making style
Multiple regression analyses were then performed to examine how well each of the three attitude factors could be predicted from the four factors relating to decision-making style. For the sample as a whole (i.e. 785 subjects with complete data), the analysis predicting 'pronuclear' attitudes yielded a multiple R of 0·16 ($F(4,780) = 5·42$, $p < 0·001$). The beta weights for the four 'style' factors were 0·12 ($p < 0·001$) for 'defensive avoidance', 0·04 (ns) for 'self-esteem/vigilance', −0·10 ($p < 0·01$) for 'passivity' and −0·03 (ns) for 'hypervigilance'. For 'involvement', the multiple R was 0·42 ($F(4,780) = 42·66, p < 0·001$), and the corresponding beta weights were respectively −0·34 ($p < 0·001$), 0·22 ($p < 0·001$), −0·11 ($p < 001$) and 0·04 (ns). For 'optimism', the multiple R was 0·18 ($F(4,780) = 6·31, p < 0·001$), and the beta weights, −0·07 (ns), −0·02 (ns), −0·00 (ns) and −0·16 ($p < 0·001$), respectively. Thus, we can infer that the measures of decision-making style were more predictive of degree of concern and involvement with nuclear issues shown by individuals than of the direction of their support or opposition to nuclear power.

Equivalent analyses on the separate groups showed a certain amount of variation. The multiple R for 'pronuclear' attitudes was significant for Adelaide (0·26, $F(4,261) = 4·74, p < 0·001$) and Exeter (0·38, $F(4,124) = 5·51, p < 0·001$) but not for the remaining nationality samples. The multiple R for 'involvement' was significant in all groups. However, that for 'optimism' was significant only for Adelaide (0·24, $F(4,261) = 4·02$, $p < 0·01$) and Aix (0·28, $F(4,146) = 3·01, p < 0·05$).

Attitudes to specific nuclear policies
As explained, the text of these items varied between national groups. Analyses were based on seven items from each version of the questionnaire. Reliability analyses were first performed to see if the different items could be summed (for each group separately) to provide an overall attitude score. The values of Cronbach's alpha were extremely high in all groups: Adelaide, 0·93; Aix, 0·89; Amsterdam, 0·92; Berlin, 0·92; Exeter, 0·93. In other words, attitudes towards various civil and military uses of nuclear power are very highly predictable from each other. Those who are opposed to military uses will also be more opposed to civil uses, and vice-versa.

Comparisons of the mean agreement scores for particular items showed significant between-item differences for each group. For Adelaide ($F(6,1734) = 87·94, p < 0·001$)

the use of Australian uranium for nuclear weapons elsewhere in the world was most rejected (1·89), but existing uranium mining in South Australia was least rejected (mean = 3·39) For Aix ($F(6,900) = 51·00$, $p < 0·001$) the installation of cruise missiles was most rejected (2·17), whereas existing French nuclear stations (3·46), and the independent French nuclear deterrent (3·95) were least rejected. For Amsterdam ($F(6,594) = 7·95$, $p < 0·001$) high-level underground waste disposal was most rejected (2·28), and existing Dutch nuclear power stations were least rejected (3·17). For Berlin ($F(6,816) = 12·06$, $p < 0·001$) the building of new nuclear power stations in East Germany was most rejected (1·34), whereas existing nuclear power stations in West Germany, though still very unpopular (1·93), were least rejected. For Exeter ($F(6,918) = 29·70$, $p < 0·001$) installation of cruise missiles was most rejected (2·61), and existing nuclear power stations in the U.K. were most tolerated (3·79). For each group, therefore, existing nuclear plants (mines in Australia) in respondents' own country were least strongly opposed.

Separate analyses were then conducted within each group to relate the attitude index (calculated by summing responses to these sets of items) to individual factor scores derived from the principal components analyses previously described. Multiple regressions to predict the attitude index from decision-making style were disappointing, only the Adelaide (0·21) and Exeter (0·27) samples yielding significant multiple Rs. Taken as a whole, therefore, this fits in with our previous finding that self-rated favourability was largely unrelated to decision-making style.

Analyses to predict the attitude index from the three attitude factors (i.e. pronuclear, economic optimism and involvement) however, were far more successful. Multiple Rs were: Adelaide, 0·74; Aix, 0·75; Amsterdam, 0·89; Berlin, 0·79; Exeter, 0·82; all $p < 0·001$. In each case, as one would expect, the beta weight for the 'pronuclear' factor was highly significant (respectively: 0·73, 0·71, 0·81, 0·67, 0·70; all $p < 0·001$). 'Economic optimism' was also a highly significant predictor in all groups (respectively: 0·14, 0·23, 0·22, 0·32, 0·28; all, $p < 0·001$). The attitude index was inversely related to 'involvement' in Adelaide ($-0·10$, $p < 0·05$) and Berlin ($-0·20$, $p < 0·001$), but not in the remaining groups.

Political preference
Broadly, these measures indicated that, in each of the five national samples, those with more conservative political views were more likely to favour nuclear power. There were fewer relationships, however, with decision-making style.

In Adelaide, those who approved more of 'the political influence of trade unions in Australia' were more antinuclear on the summated index ($r = -0·19$, $p < 0·001$), had more 'antinuclear' factor scores ($r = -0·22$, $p < 0·01$), and were more economically optimistic ($r = 0·16$, $p < 0·01$). They were also lower on 'defensive avoidance' ($r = -0·17$, $p < 0·01$).

Similar correlational analyses were performed on the Berlin data, where three items could be taken as indicative of political orientation. A more pronuclear attitude index was associated with more approval of government planning for nuclear catastrophes ($r = 0·34$, $p < 0·01$), and with disapproval of the influence of the environmentalist ($r = -0·38$, $p < 0·001$) and peace movements ($r = -0·54$, $p < 0·001$). The same was true for the 'pronuclear' factor score (rs = 0·24, $p < 0·01$; $-0·56$, $p < 0·001$; $-0·51$, $p < 0·001$), and for 'economic optimism' (rs = 0·39, $p < 0·001$; $-0·15$, $p < 0·1$; $-0·37$, $p < 0·001$).

The remaining groups were asked more directly to indicate their preferred political party. In Aix, strongly pronuclear attitude indices were given by three members of the far-right Front National, these were followed by the right/centre-right Rassemblement pour la Republique (RPR) and the Union pour la Democratie Francaise (UDF), then the Socialists, then the Communists, and then the Ecologists ($F(5,106) = 11\cdot90$, $p < 0\cdot001$). Exactly the same ordering applied to the 'pronuclear' factor score ($F(5,106) = 7\cdot40$, $< 0\cdot001$). The Front National supporters scored highest, and the RPR supporters lowest, on 'involvement' ($F(5,106) = 2\cdot85, p < 0\cdot05$). On 'economic optimism' the Communists scored lowest, followed by the Socialists ($F(5,106) = 4\cdot09$, $p < 0\cdot01$). With regard to the 'style' factors, Communists and Socialists scored lowest on 'defensive avoidance' ($F(5,106) = 3\cdot67$, $p < 0\cdot01$), but highest on 'hypervigilance' ($F(5,106) = 2\cdot89$, $p < 0\cdot05$).

In Amsterdam, a similar straight right-to-left ordering (Conservatives, Christian Democrats, Social Democrats, Socialists, Greens) was found for both the summated attitude index ($F(4,88) = 49\cdot74$, $p < 0\cdot001$) and the 'pronuclear' factor score ($F(4,88) = 27\cdot07$, $p < 0\cdot001$). In Exeter, Conservatives were most pronuclear, followed by Liberal Social Democrat (SDP) supporters and then Labour supporters, the latter groups being closer together on both the summated attitude index ($F(2,106) = 29\cdot47$, $p < 0\cdot001$) and the 'pronuclear' factor score ($F(2,106) = 19\cdot91$, $p < 0\cdot001$). The same ordering applied to 'economic optimism' ($F(2,106) = 18\cdot93$, $p < 0\cdot001$). Labour supporters were highest, and Liberal/SDP supporters lowest on 'involvement', ($F(2,106) = 6\cdot59$, $p < 0\cdot01$), the reverse being true for 'defensive avoidance' ($F(2,106) = 10\cdot72$, $p < 0\cdot001$).

Discussion

The main conclusion from our results is that interpretations of the seriousness of the Chernobyl accident by subjects in all five national samples were closely related to their attitudes on *other* nuclear-related topics, for instance waste-disposal and nuclear deterrence. This is strongly suggestive of a process of assimilating new information to pre-existing schemata (of nuclear power as a whole being 'bad' or 'not so bad'). Such an assimilative process may lessen the absolute size of any changes in attitude that events such as the Chernobyl accident may produce (Eiser *et al.*, 1989).

Individual differences in political decision-making style were identified, but predicted favourability of nuclear attitudes only for the English-speaking samples. A possible interpretation is that the decision-making items lost some of their sensitivity in translation. On the other hand, levels of 'involvement' were consistently related to coping style overall and within each national group considered separately. Those individuals who were more 'involved' (who paid more attention to, and were more fearful of the accident) were significantly lower on the style of 'defensive avoidance' (except in Aix) and significantly higher on that of 'self-esteem/vigilance' (except in Amsterdam). These findings accord with our general hypothesis that differences in patterns of decision-making relate to the degree of anxiety and attention evoked by the Chernobyl accident. There is also support for Janis and Mann's (1977) analysis of defensive avoidance and vigilance as contrasting means of coping with threatening information.

The differences between the various national samples should be interpreted cautiously. All were 'convenience' rather than representative samples, although the manner of their recruitment was comparable. Even so, a number of the more obvious differences, such as the lower 'involvement' of the Australians, and the strength of antinuclear feeling in Berlin, make sense in terms of geographical and (possibly) broader socio-political considerations. However, we were primarily looking for consistencies across samples rather than differences between them. One of the most striking of these consistencies is the relationship between more pronuclear attitudes and more right-wing political preferences, even granted the varied ways in which such preferences were assessed.

All this suggests that attitude differences on nuclear issues form part of far broader value systems and ideologies. Such differences are unlikely to be settled by mere factual information, suggesting that specific costs or benefits have been over- or under-estimated, as the very relevance or salience of such costs or benefits will be perceived differently by those with different attitudes (van der Pligt, Eiser & Spears, 1986b). The Chernobyl accident was perceived as serious and threatening, but to different extents and in different ways depending on individuals' broader attitudes and decision-making style.

References

Abelson, R. P. (1959). Modes of resolution of belief dilemmas. *Journal of Conflict Resolution*, **3**, 343–352.

Baum, A., Gatchel, R. J. & Shaeffer, M. A. (1983). Emotional, behavioral and physiological effects of chronic stress at Three Mile Island. *Journal of Community and Clinical Psychology*, **51**, 565–572.

Burnett, P. C., Mann, L. & Beswick, G. (1988). Validation of the Flinders Decision Making questionnaires in course decisions by students. Unpublished manuscript. Flinder University of South Australia.

Crocker, J., Fiske, S. T. & Taylor, S. E. (1984). Schematic bases of belief change. In J. R. Eiser, Ed., *Attitudinal Judgment*. New York: Springer-Verlag.

Eiser, J. R., Spears, R. & Webley, P. (1989). Nuclear attitudes before and after Chernobyl: Change and judgment. *Journal of Applied Social Psychology*, **19**, 689–700.

Eiser, J. R., Spears, R., Webley, P. & van der Pligt, J. (1988). Local residents' attitudes to oil and nuclear developments. *Social Behaviour*, **3**, 237–253.

Festinger, L., Riecken, H. & Schachter (1956). *When Prophecy Fails*. Minneapolis: University of Minnesota Press.

Janis, I. L. (1951). *Air War and Emotional Stress*. New York: McGraw-Hill.

Janis, I. L. (1962). Psychological effects of warnings. In G. W. Baker & D. W. Chapman, Eds., *Man and Society in Disaster*. New York: Basic Books.

Janis, I. L. & Mann, L. (1977). *Decision Making: a Psychological Analysis of Conflict, Choice, and Commitment*. New York: Free Press.

Janis, I. & Rausch, C. (1970). Selective interest in communications that could arouse decisional conflict: A field study of participants in the draft-resistance movement. *Journal of Personality & Social Psychology*, **14**, 46–54.

Mann, L. (1982). *Decision Making Questionnaires I and II*. Unpublished Scales. Flinders Decision Workshops, The Flinders University of South Australia.

Nealey, S. M., Melber, B. D. & Rankin, W. L. (1983). *Public Opinion and Nuclear Energy*. Lexington, MA: Lexington Books.

Radford, M., Mann, L. & Kalucy, R. (1986). Psychiatric disturbance and decision making. *Australian and New Zealand Journal of Psychiatry*, **20**, 210–217.

Van der Pligt, J., Eiser, J. R. & Spears, R. (1986a). Attitudes toward nuclear energy: Familiarity and salience. *Environment and Behavior*, **18**, 75–93.

Van der Pligt, J., Eiser, J. R. & Spears, R. (1986b). Construction of a nuclear power station in one's locality: Attitudes and salience. *Basic and Applied Social Psychology*, **7**, 1–15.

Webley, P., Eiser, J. R. & Spears, R. (1988). Inflationary expectations and policy preferences. *Economics Letters*, **28**, 239–241.

Webley, P. & Spears, R. (1986). Economic preferences and inflationary expectations. *Journal of Economic Psychology*, **7**, 359–369.

PUBLIC RESPONSES TO THE CHERNOBYL ACCIDENT

ORTWIN RENN

Center for Technology, Environment, and Development (CENTED),
Clark University, 950 Main Street, Worcester, MA 01610, USA

Abstract

The reactor accident at Chernobyl caught many European nations by surprise since most risk management institutions were unprepared for an accident of the magnitude and transnational character of Chernobyl. Although confusion and contradictory advice from these institutions dominated the risk management efforts in the early aftermath of the disaster, the dose savings achieved by protective actions were roughly proportional to the magnitude of the nuclear threat. The accident itself and the policies adopted to cope with the fallout had a major effect on public opinion. This effect was the more dramatic and enduring, the more a country was affected by the fallout and the higher the percentage of indifferent positions toward nuclear power was prior to the accident. The media certainly intensified public concern, but did not distort the seriousness of the risk or create confusion about what protection actions were adequate. The major lesson from the disaster is to have a better risk management and communication program in place before a disaster strikes.

Public Responses to the Chernobyl Accident

The reactor accident at Chernobyl in the Soviet Ukraine on April 25–26, 1986, posed a threat of radioactive contamination to various countries. The event provided a unique experience for studying attitude changes and public reactions to an identical stimulus in a multi-cultural context. Since the risk management institutions of most affected countries were unprepared for an accident of the magnitude and transnational character of Chernobyl, it was necessary to improvise appropriate responses to the fallout. A particular problem was the highly nonuniform distribution of ground deposition produced by rainout (Hohenemser *et al.*, 1986).

Both the disaster and the attempts to cope with the fallout situation had significant effects on public opinion and attitudes in the countries involved. Public responses were not only directed toward the origin of the disaster, the Chernobyl Nuclear Power Plant, but included general considerations about the acceptability of nuclear power, the domestic nuclear power program, and the efficacy of risk management in each country. Although the official positions of each country's government toward nuclear energy were rarely reversed after the accident (Flavin, 1987), public opinion shifted towards a more skeptical overall position towards nuclear power, in almost all affected countries. Opposition parties in West Germany and the United Kingdom responded with a decision to phase out nuclear energy if they were given the opportunity to take over the government.

The focus of this article is on four interrelated subjects: first, the official responses of risk management agencies and national governments are briefly described to provide a platform for the discussion of the public responses; second, the shift of public opinion in many affected countries is presented and discussed: third, attitude change and behavioral responses are examined and attempts are made to explain them by referring to attitude theory, institutional credibility, and mass media influence; and fourth, lessons for risk communication and management are formulated to reflect the results of our investigations of public reactions. Our analysis includes only the West European countries and (partially) the United States.

Official responses of European governments to Chernobyl
European risk management institutions faced a serious crisis when the fallout of the Chernobyl accident continued to travel from the East to the West. Most countries had no emergency plans for coping with accidents that occurred outside of their territory or outside the European community. Common standards were missing, and the ALARA (As Low As Reasonably Achievable) principle was often too vague to make fast, effective, and consistent decisions. In addition, the following factors aggravated the situation (Renn, 1986):

- lack of capability for food monitoring;
- lack of risk communication programs about the nature of the health effects from low dose radiation;
- problems of justifying different protective actions for different regions; and
- control of public responses (compliance with public recommendations, but avoidance of overreactions).

Most observers agree that the European countries had difficulties in overcoming these management problems. They apparently failed in granting optimal protection for the population at risk as well as in assuring the public that a clear and consistent management approach was taken (Otway *et al.*, 1987; Wallmann, 1987). The confusion was heightened by the inconsistent use of units of measurement, the politicization of the issue by specific interest groups (for example, environmentalists and the nuclear industry), the public fear of radiation, and overlapping responsibilities (Otway *et al.*, 1987; Roser, 1988).

As a case in point, on the West German side of Lake Constance dairy cattle were kept off pastures, and iodine milk levels peaked at around 100 Bequerel per liter (Bq/l); whereas on the Swiss side, cattle grazed on the fresh fallout, and iodine milk levels peaked at around 1000 Bq/l (Hohenemser & Renn, 1988). Using a range of national reports, the general structure of protective actions for both eastern and western countries has been surveyed by the U.S. Interlaboratory Task Force (DOE, 1987). A similar, even more detailed data base has been prepared for OECD countries (NEA, 1988). The following list contains the most common measures and/or recommendations.

- informational activities, such as education about potential health effects and precautionary actions;
- recommendation to restrict outdoor activities such as keeping children indoors during rainfall; close swimming pools, playgrounds, and other public recreational facilities; and cancelling sport or other outdoor activities and events;
- measures to limit the ingestive pathway via controls on (1) rainwater for drinking by people and domestic animals; (2) open grazing by dairy cattle; (3) marketing of

milk and dairy products; (4) banning of imported foodstuff; (5) washing of leafy
vegetables; and (6) controlling meat sources (domestic cattle and game);
- efforts to limit external exposure via evacuation, sheltering, and administration of
  iodide pills (only Soviet Union);
- environmental controls, including requirements for changing industrial and
  hospital air filters and controls on the use of sewage sludge for soil amendment;
- compensation for agricultural losses.

How effective were these protective actions that have been or are still in effect in
various countries? According to the U.S. Interlaboratory Task Force (DOE, 1987), the
question cannot be usefully answered because the pattern of application is
insufficiently clear, and the effect of future actions cannot be ascertained. At the same
time, given the eight-day nuclear halflife of iodine 131 and the one-year halflife of
cesium 137 in most food chains, one would expect most dose savings to occur in the first
year. The Nuclear Energy Agency of the OECD (NEA, 1988) has asked member
countries to estimate first year dose savings, with results shown in Table 1.

It is clear from the Table that collective dose savings varied widely. Hohenemser and
Renn conducted a regression analysis to relate dose savings to concentration of
radionuclides (Hohenemser & Renn, 1988). In spite of the small data base, the
correlation factor between dose savings and whole body dose is quite substantial. The
authors calculated a correlation coefficient of 0·70.

The relationship between dose and dose saving indicates that most West-European
nations acted in proportion to the nuclear threat. The collected data reflect, however,
only national averages for both dose and dose reductions. Additional studies are
needed to relate regional dispersion of radionuclides with dose savings. Anectodal
evidence collected in Germany (Renn, 1986; Roser, 1988) and in France, Italy, and the

TABLE 1[a]

Estimates of first-year whole body dose savings from protective actions (in percent)

Country	Collective dose saving			
	Total	Infants	Children	Adults
Austria	50	53	50	50
Belgium		——————— very small ———————		
Finland	7·2	12	11	6·3
France		——————— very small ———————		
West Germany	30	(Overall estimate)		
Greece	23	25	17	24
Italy	18	53	33	15
Luxembourg	7·5	17	13	6·6
The Netherlands	15	43	23	12
Norway	32	29	28	33
Sweden	15	0	3	17
Switzerland	1	50	0	0
Turkey	12	0	18	11
United Kingdom	1	<1	1	1

[a] Source: Nuclear Energy Agency, The Radiological Impact of the Chernobyl Accident in OECD Countries
(Paris: NEA 1988).

United Kingdom (Otway *et al.*, 1987) suggest that the allocation of dose reduction measures to local hot spots was insufficient and often counterproductive.

Policy adjustments after Chernobyl

In addition to management responses, most countries adopted policy changes in their domestic nuclear program. Soon after the accident, the Dutch parliament approved a motion to suspend a decision on the location of two nuclear reactors until a thorough analysis and evaluation of the Chernobyl accident had been completed (*Nucleonic Week*, 1986). In Yugoslavia, the Croatian parliament voted to reappraise the Prevlaka nuclear power plant. Sweden reaffirmed its national policy of terminating nuclear energy in the future and Austria confirmed its decision not to use nuclear energy.

West Germany reacted with the setting up of a Federal Ministry for Environment and Reactor Safety. Finland postponed new orders for nuclear power plants. In Italy and Switzerland, a new petition for banning nuclear power was initiated. The Italian referendum was largely accepted although the impacts of this referendum are still unclear; the Swiss referendum to ban nuclear power there was defeated (WEC, 1989).

It needs to be noted that these political actions did not change the basic nuclear policies in most countries. Those countries already determined to phase out nuclear energy or not to use nuclear energy at all confirmed their decision, while countries with an ongoing nuclear program continued to support nuclear energy, but slowed down expansion of the program and initiated thorough reviews of the existing safety concepts (Flavin, 1987). Greece decided not to pursue further the nuclear option as a result of Chernobyl.

The confusion caused by the lack of intervention levels and inconsistent use of existing standards in the aftermath of Chernobyl led to increased efforts of most European countries to define and promulgate secondary intervention levels on a national and international scale. In late June 1986, the European Community adopted new common standards for cesium, a similar agreement could not be reached for iodine; France and the United Kingdom opted for a more lenient level than West Germany, The Netherlands, Italy, and other members of the EC (Hohenemser & Renn, 1988).

In addition to specifying intervention levels, which may not be exceeded but may be set lower than prescribed (ALARA), the German parliament passed a new act on the precautionary protection of the population against radiation exposure in December 1986 (Gesetz, 1986). The major goal of the new act is to ensure that a comprehensive monitoring system is in effect for the whole country and that measurement and intervention levels are unified throughout the eleven German states. Since the ALARA principle has also been confirmed as the basic response philosophy in section 1 of the new act, each state has the flexibility to go beyond federal standards. The collection and interpretation of data are clearly the responsibility of federal agencies, but the selection of actions is still state responsibility unless federally determined dose limits are exceeded (SSK, 1987).

Public attitudes after Chernobyl

Opinion polls to investigate the public attitude towards nuclear energy were conducted in almost every country after the accident (*Newsweek*, 1986; *Wall Street Journal*, 1986; Flavin, 1987; Allensbach, 1987; Suhonen & Virtanen, 1987; Midden & Verplanken,

1990). Not surprisingly, support for nuclear power declined in most countries and in spite of some recent recovery has not reached the level of approval of the pre-Chernobyl period. Opposition to nuclear power immediately after the accident peaked in Finland, Yugoslavia, and Greece (over 30%); considerable increases (over 20%) were observed in Austria, West Germany, and Italy; moderate changes took place in the United Kingdom, France, The Netherlands, Sweden, and Spain (12–18%).

Different survey results suggest that public response was aggravated by poor risk communication (Otway *et al.*, 1987). Frequently, citizens were convinced that the government was withholding information and did not tell the truth (63% of the French population, for example). In Germany, well-educated citizens complained that the government did not give enough and adequate information, while less educated citizens felt overwhelmed by the flood of information and opted for more consistent and understandable messages (Peters *et al.*, 1987; Peters, Albrecht, Hennen & Stegelmann, 1990).

In the U.S.A. which was scarcely affected by the fallout, public opposition to nuclear power gained another 5% to reach a peak of 49%, the highest percentage ever reported. After the accident at TMI, public support was higher than in the aftermath of Chernobyl (*Newsweek*, 1986). An even more dramatic change was that the opposition to a nuclear plant in the respondent's neighborhood increased from 45% in 1976 to 60% in 1979 and 70% in May 1986. Recent polls confirm, however, that the level of opposition has almost fallen back to the pre-Chernobyl level (WEC, 1989).

Observers of East-European countries detected a growing opposition to nuclear power in Poland, Hungary, and specifically CSSR (*Nucleonic Week*, 1986, 1987). Data are only available for Yugoslavia. Three months after the accident the number of opponents towards nuclear energy increased from 40 to 74%. One year after the accident the number of opponents (64%) is still 24% higher than before the accident.

After the initial shock, many supporters of nuclear power who had expressed negative attitudes in the immediate aftermath of Chernobyl changed their opinion again during the months following the accident and regained their initial positive attitude. Figure 1 shows the increase of opposition measured two to three months after the accident, and at least one year after the accident. A considerable decline in public opposition has occurred in each country where data were available. The new numbers are almost proportional to the distribution of opponents observed directly after the accident. Greece and Yugoslavia are still leading the list of countries with the most dramatic changes in attitude; Germany and Italy are in the middle positions, and in France and the United Kingdom, the changes were the least enduring.

The situation is even more complex in the two Scandinavian countries of Sweden and Finland. Two-thirds of the Finnish population who rejected nuclear power as a means of energy production in the aftermath of Chernobyl, changed their opinion within a year and became undecided or even supportive of nuclear power. A monthly poll taken in Finland reveals that the peak of the opposition was reached in summer and early fall 1986, but that opposition declined rapidly in winter 1986/1987 (Suhonen & Virtanen, 1987). In Sweden, the extended discussion on contaminated game and moose meat triggered a new wave of antinuclear opinion more than 14 months after the accident (WEC, 1989).

In spite of the recovery of public support in most countries, all available data clearly demonstrate that the initial distribution of positions towards nuclear energy has not

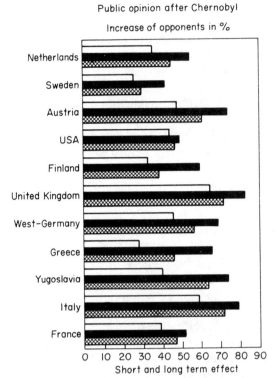

FIGURE 1. The changes of public opinion before, directly after, and one year after the Chernobyl accident in selected countries. ▨ = 1987; ■ = after; ☐ = before.*

* Sources: Flavin 1987; Allensbach 1987; Roser 1988; Suhonen & Virtanen 1987; Otway *et al.*, 1987.

been reached in any country. One year after the accident it is not clear whether nuclear attitudes will remain negative, or may change again—possibly in both directions.

Explanations for attitude changes

To gain a better understanding of long-term changes in public attitudes, it is interesting to compare the responses after Chernobyl with those after TMI. The accident at TMI proved to have only a limited effect on public opinion in those countries that were not directly affected by the consequences of the accident and which had a low percentage of undecided positions or 'don't know' responses prior to the accident (Renn, 1984). While US public attitudes after TMI became more critical and remained that way thereafter, Germany, France, Great Britain, and most Scandinavian countries experienced only a short period of increased opposition. Within a year, public confidence in nuclear energy reached pre-TMI levels or was actually even higher. In contrast the population of countries such as Spain and Italy which had a high number of 'don't knows' and uncommitted opinions at that time, became more skeptical about nuclear power and consolidated this skeptical orientation over the following years.

A theoretical explanation for this behavior may be found in the 'inoculation effect' of attitude formation and commitment (McGuire, 1985; Renn, 1984). This effect makes individuals with a positive attitude feel amost immunized against negative incidents, while an uncommitted person may use the incident as an incentive to take a side in the debate. The metaphor of inoculation refers to the preparedness of the immunity system to cope with a class of bacteria or virus. Similarly, the mental system tries to avoid the exposure or the storage of information that would induce painful changes of previous attitudes (Cotton, 1985; Frey, 1986). Selective exposure and downplaying of counter-evidence are two mechanisms of avoiding cognitive dissonance (Festinger, 1957).

Proponents of nuclear energy who have been inoculated with information about minor incidences in nuclear facilities, react first with the shock response of withdrawing their support, but consolidate their original attitude after a while. This is particularly true for these persons for whom nuclear energy has been a peripheral subject, i.e. a subject or issue that evokes low personal involvement and relevance. As shown, attitudes about peripheral issues tend to be based on heuristic processing and simplified reasoning (Chaiken, 1980; Petty & Cacioppo, 1986). Arguments and content related reasoning, however, are less prevalent for peripheral issues. If a 'psychologically undeniable' accident occurs, arguments to cope with this event are missing under this route of attitude formation. Instead cues, such as opinions of reference groups, trust in institutions, the alleged viewpoints of the majority, become important orientations in attitude formation and change. These cues are usually negative immediately after the accident, hence the impossibility to find reassurance of existing (positive) attitudes. Over time, professional groups and interest groups provide new positive cues so that the initial shock can be overcome. Such a recovery to the original position requires, however, the existence of counter-evidence, based on reassuring cues, such as a perceived positive opinion of the majority or of highly esteemed reference groups, assurance that the incident was not as severe as originally anticipated, and messages from highly trusted sources that similar events would not reoccur or could not happen in one's native country.

People that regard nuclear energy as a central issue have usually collected enough arguments and are able to cope with large accidents as long as these accidents are not proving the inaccuracy of the previously held beliefs. Transitional cues, such as the immediate response of politicians or interest groups, have less impact on their attitude. But if salient beliefs are at stake, for example, about the nature of nuclear safety, changes of attitudes may occur and are probably stable unless new information becomes available that reassures the initial beliefs. Due to the inoculation effect and previous commitments to a pro-nuclear attitude, such changes are less likely to happen and may extend over a longer period of time. Attitudes may slowly erode over time if salient beliefs are consistently challenged.

The data collected for West-Europe and Yugoslavia allow a quasi-empirical test of the concept of inoculation. If this effect had dominated the response pattern after Chernobyl, the increase in opposition should be a function of two factors: the degree to which people perceive the incidence as counterproving salient beliefs (overcoming the inoculation effect), and the quantity of indifferent and 'don't know' responses towards the issue in question (absence of inoculation effect). Since we lack data measuring the perception of sufficient evidence to change salient beliefs, we hypothesized that such evidence would be proportional to the real danger posed by the incident. The more people are forced to face a health threat of an object associated with favorable attitudes, the more likely it is that salient beliefs would be affected. Therefore, increase in opposition should be proportional to the radiological doses experienced in each country and also be correlated with the number of indifferent positions before the accident.

The correlation between whole body doses and the increases of public opposition is indeed substantial. The correlation is 0·82 ($p < 0·01$); the rank order coefficient is 0·79 ($p < 0·01$). The increase of opponents towards nuclear energy is therefore directly proportional to the released body doses. Furthermore, the correlation between the number of indifferent positions before and the number of opponents after the Chernobyl accident is also fairly high. The coefficient R is 0·42 ($p < 0·20$); the rank order coefficient is 0·54 ($p < 0·10$). The lower correlation coefficients are basically due to the lack of variation in the independent variable and the small sample size; in most West-European countries, less than 5% had indifferent of 'don't know' positions towards nuclear energy before the Chernobyl accident.

Furthermore, inoculation theory would predict an almost complete recovery of public opinion in countries with strong prior commitments to a specific attitude and weak stimuli to change salient beliefs. Likewise we expect hardly any recovery in countries with a low degree of attitudinal commitment, and the presence of strong stimuli to change salient beliefs. Again we correlated the recovery effect, measured in terms of percent opponents one year after the accident, with the dose received after Chernobyl. The correlation coefficient is almost identical with the one using the number of opponents immediately after the accident. The correlation is 0·79 ($p < 0·01$) and 0·78 for the rank order correlation. A similar result was obtained for the number of indifferent positions (Spearman = $-0·52$).

What long-term effects can we then expect from the Chernobyl accident taking into account the implications of the 'inoculation' concept? The studies undertaken so far about the effects of the Chernobyl accident imply the following pattern: those countries less affected by the fallout and with highly structured attitudes prior to the accident, such as France, the United Kingdom and Spain, exhibited less public concern from the beginning, and within one year public support towards nuclear energy almost reached pre-Chernobyl levels.

West Germany, The Netherlands, Austria, Switzerland, Sweden and Italy were more affected by the fallout, but had highly structured attitudes prior to the Chernobyl accident. Therefore, the initial responses in those countries were much more dramatic than in France, for example, but a significant recovery of pro-nuclear attitudes occurred within the first year after the accident. Finland is probably the best example for the inoculation effect where the initial shock resulted in more than 30% additional opponents from which two-thirds revised their opinion again during the following year (Suhonen & Virtanen, 1987). With continuing negative cues, however, persons with a peripheral interest in nuclear power are more inclined to form more negative beliefs about nuclear power and to change their overall attitude.

Another interesting case is Greece. This country, although less affected, had a strong proportion of uncommitted or undecided positions in the nuclear debate. As a result, public opinion changed dramatically and the recovery was only marginal compared with Finland (Otway et al., 1987).

If this pattern prevails, it suggests that revealed attitude changes will only last in those countries which either experienced impacts of the accident and/or had a strong proportion of uncommitted positions prior to the accident. This would mean that public attitudes in the United Kingdom, France, Spain, Ireland, and the U.S.A. will not be strongly affected by the Chernobyl accident in the long run. The attitude changes in Central and Northern European countries may further recover, but will not reach the pre-Chenobyl level (similar to the TMI experiences in the U.S.A.). The opposition to nuclear power will, therefore, gain influence in these countries, but may not be able to form a majority. The most dramatic changes are to be expected in countries in which both conditions were met, in particular Poland, Hungary, the CSSR, and Yugoslavia. Except for Yugoslavia, survey data are not available to test this claim.

Protective actions by individuals
Given all the confusion about protective actions and the changes in public opinion, it is not surprising that many people overreacted and others did not even follow the simplest recommendations. Otway *et al.* (1987) report the following spontaneous reactions that often led to a higher risk to the individual than the one to be avoided:

- a sudden increase in the number of abortions (reported in Austria and Italy);
- panic buying of tinned, frozen, and other long-life foods, reported in most countries, but reaching near-riot proportions in Greece;
- buying of radiation measuring equipment for personal use (reported in West Germany and Great Britain);
- intake of potassium iodide (sometimes in large overdoses), reported in Poland, West Germany, and Denmark; and
- an increase in suicides partly attributed to inability to cope with the threat, partly attributed to the financial ruin of small firms (reported in Italy and Greece).

German newspapers reported other types of responses, such as removal of topsoil in private gardens, staying indoors for almost two weeks, and burning of clothes worn while there was fallout. Although such overreactions received quite a strong press coverage, they were not at all typical of the majority of the population. As Peters *et al.* have shown in their study (1990) most respondents in a West Germany survey did not engage in any protective action. In their study, 55% of the respondents declared that they had not changed their diet after Chernobyl.

A minority of the population, in general better educated and more aware of environmental problems than the average citizen, was extremely worried by the fallout from Chernobyl and responded with corresponding protective actions highly publicized in the media. This minority perceived nuclear energy as a central issue and was convinced that major actions were necessary to cope with the risk. The majority reacted with much more apathy and did not perceive an immediate need for self-protection. The anger and frustration resulting from confusion and uncertainty about the adequacy of personal reactions were partly channelled towards the object that appeared to have caused all the upset. That is why attitudes towards nuclear power were much more affected by the accident than was personal behavior.

Trust in emergency institutions and information
Another possible factor that may explain the low compliance rate with official recommendations is the loss of trust and belief in government institutions. For West Germany and the United States, survey results have yielded a clear correlation between attitudes towards nuclear power and confidence in public institutions (Renn, 1984). The study by Peters *et al.* revealed again an astonishing result for West Germany. In spite of the confusion and contradictions created by the official emergency managers, around 60% of all respondents indicated that they found the federal government and

other official institutions, such as the nuclear research centers, totally, or at least partially, trustworthy.

This result may be typical only for Germany since public opinion polls in Italy and France reported 70% or more of respondents feeling distrust towards and lack of confidence in the government (Otway et al., 1987). Trust in government was not high in these two countries before Chernobyl, and their rather restrictive handling of information may have aggravated this feeling. Although survey data are not yet available, the impressions gained in Sweden and The Netherlands support the notion that there was sufficient confidence in the emergency handling institutions.

The German data suggest that most people had equal confidence in pronuclear and antinuclear institutions (Peters et al., 1987). The (pronuclear) government and the nuclear research centers were trusted to nearly the same degree as their direct adversaries, the (antinuclear) citizens' initiatives and the ecological research institutes. Furthermore, those who trust the pronuclear information sources, do not necessarily mistrust the antinuclear sources.

A large percentage of the population appears to believe that the often contradictory elements of information given by both camps contain a certain amount of truth, and that both sides do not lie deliberately but focus on those aspects that support their general viewpoint. Clearly, confidence in two antagonistic camps add to the confusion experienced in the aftermath of Chernobyl, and caused frustrations from listening to seemingly contradictory sets of recommendations from respected institutions.

The politicalization of the emergency response arena not only added to the confusion of the public, but destroyed the potential role of scientific institutions as arbiters in the conflict between the major interest groups. Scientific institutions proved unable to provide unbiased evidence for orientation and suggestions for effective hazard management. It is doubtful, however, whether the politicization of scientific institutions can ever be reversed.

The role of the media

The media played a major role in amplifying the dissent among science institutions and in taking part in speculation about potential health effects. The coverage of over-reaction and the emphasis on inconsistencies between official recommendations were additional causes of public discomfort and skepticism. But all these phenomena were real and not invented by the media. Overlapping responsibilities, contradictive advice, inability to explain the meaning of specific countermeasures, and total chaos in the units and intervention levels characterized the European scene. Media coverage was merely a reflection of what actually happened in most countries.

Content analyses of the media were conducted in many European countries. In Italy articles in print media were the focus of an investigation on information content, information matter, and expressed biases (Belelli, 1988). The results confirm that the press did not dramatize the effects of Chernobyl and did not distort the factual evidence. Although it may be difficult to assess, correctness of information was evaluated by the investigators and the results compared with the pre-Chernobyl situation.

The basic opinion on nuclear energy that each print medium developed prior to the Chernobyl accident were in general not changed after the accident. A little more than half (50·3%) of all Italian newspapers were in favor of nuclear energy before Chernobyl. This number dropped to 48·7% after Chernobyl. This may also be seen as an indication of the inoculation effect, this time exerting its influence on an institution rather than an individual.

The most frequently reported subject in Italy's print media was not the accident and its causes and consequences, but the reflections on the domestic nuclear program (25% of all analysed articles). Similar results were yielded in a German and a Finnish study (Rager, 1987; Joutsenniemi, 1987). The attention of the media shifted from the accident to the potential danger of domestic nuclear reactors. Most articles or TV reports discussed the future of the nuclear industry in each country. The change of media topics ran parallel to the shift of concern in the general public: the public was less worried about the fallout and the demanded changes in diet, but was concerned about the adequacy and political justification of the domestic nuclear program.

In West Germany, the media coverage peaked on the question of the possibility of a slow or fast phase-out of nuclear energy. Again, it should be emphasized that the discussion in the media was predominantly a reflection of the argumentation among social groups and political parties (in West Germany between the Christian Democrats advocating further use of nuclear energy and the Social Democrats advocating a phaseout within ten years). The media did not provoke this discussion or initiate it. In addition, most German media tried to give justice to both sides of the nuclear debate. Proponents of nuclear energy were just as often cited as opponents (Rager, 1987).

With the exception of the two major TV programs (ARD and ZDF), however, which were well-balanced in presenting pronuclear and antinuclear arguments in their coverage of the Chernobyl accident and its domestic consequences, the print media usually took clear position if favor of or against nuclear energy. Most of the print media ranging from the liberal to the left political spectrum adopted an antinuclear viewpoint, whereas the conservative media, such as 'Die Welt' or 'Frankfurter Allgemeine' voiced a positive opinion towards nuclear energy. The media's points of view correspond, therefore, with the political programs of the conservative and progressive parties, although the liberal party had not finally decided on its policy towards nuclear power. The positions in the parties, as well as in the media, were not initiated through Chernobyl but sharpened and reinforced.

The results of the Finnish study point in the same direction. Management problems concerning the domestic program were the major issues in the print media (Joutsenniemi, 1987). Antinuclear positions were reconfirmed and pronounced more vigorously, but also pronuclear media criticized the government for mismanagement of the crisis and the creation of public confusion.

A media analysis for seven different countries based on print media comes to a similar conclusion (Otway *et al.*, 1987): The coverage was in general accurate and reflected the political debate in each country. Furthermore, the authors revealed that there had been reasonably fair coverage of the event considering the time constraints under which journalists usually operate. Another example of the willingness of the press to cooperate with scientists and to serve public interest was reported in southern Germany. A close cooperation between the scientists of the university here and the local newspaper was established in Constanz which led to an exceptionally well-organized risk communication and management effort (Hohenemser *et al.*, 1986).

The overall impression of a responsible, accurate, and fair coverage of the event by the media revealed by content analysis can be further substantiated by the subjective perception of the press coverage through public opinion. The study of Peters *et al.* (1987) indicated that 43% of all respondents felt that the electronic media covered the incident and its consequences in a well-balanced and unbiased way; 14% regarded the coverage as too negative, 12% as too positive. Those who detected biases in the media were more likely to be extreme proponents or opponents of nuclear energy, who usually reject anything as biased which does not support their point of view. Once again, the high number of 'don't know' responses is interesting, further proof of the uncertainty created by the official risk communicators.

The common prejudice, that the media were responsible for the confusion and the difficulty to convey the true dimension of the health threat, has to be rejected in the light of the evidence presented here. The media may have aggravated the feeling of confusion and frustration by pointing to the weakness of risk management and the overt contradictions in risk communication. But they only amplified what actually happened in the attempt of government officials to master the crisis. The deficiency of management response was, therefore, neither due to biased media coverage nor as analysed earlier to the lack of confidence in management institutions. It was basically a product of the malperformance of the management institutions.

Lessons for risk management and communication
The bottom line of the risk management and communication efforts in most European countries is that the degree of protective actions were generally proportional to the threat posed by the radioactive fallout, but that inconsistencies in risk management, downplaying and dramatizing by stakeholder groups, insufficient attempts to provide accurate, unbiased, and effective risk information, and helplessness vis-a-vis over-reactions and apathy of public groups all led to suboptimal results and an even more dramatic loss of reputation among the public. Confusion and inconsistencies created a climate of insecurity in which over-reactions and apathy developed simultaneously, depending on the saliency of the nuclear threat to the affected person. Many persons reacted with anger and outrage, and channelled their dissatisfaction with the risk management efforts to an increase in opposition towards the domestic nuclear program.

The Chernobyl accident is a perfect case study of the kind of mistakes and problems in nuclear emergency situations that should be avoided in the future. In fact, the lessons drawn from this case can also be applied to other supranational disasters such as chemical spills. By reviewing the first analyses of the communication and management efforts by European governments and comparing these results with some guidelines developed in the context of risk communication, the following possible improvements in future emergency responses are suggested (Renn, 1988):

(1) The unexpected high variation in local exposure to radiation requires monitoring capability on the community or district level. In order to keep costs low, regional universities or schools of higher education should be equipped with measuring devices and exact guidelines on how to use them. If there is a nuclear emergency, the designated teacher or professor, assisted by students, would be expected to measure radiation levels in the local area. Soil, water, rainfall and food should be investigated. All measurements would have to be reported to the appropriate emergency management institution according to a prearranged plan.

(2) Within each country or, even better, within the European Community or other international bodies (perhaps the IAEA), standards for radiation levels that require protective actions should be determined and promulgated. These universal standards

should serve three functions: to facilitate the responses of regional management institution by providing clear and unequivocal instructions; to ensure that no group of citizens be exposed to greater risk than others (equity concern); and to convey an easily understandable and comprehensible safety and protection rationale to affected citizens. The threshold for initiating actions should be low enough to convince the public that a sufficient degree of protection can indeed be achieved. Although the ALARA principle would allow different standards for different situations, I believe that consistency and clarity are superior goals to hazard management than cost-effective risk minimization.

(3) The type of protective action should correspond to the activity level specified for intervention and tailored to the nature of the local environment. Although a single set of protective actions for each intervention point would be most desirable from the point of view of risk communication, local circumstances and the specific characteristics of the affected region are likely to require differentiated sets of protective actions based, however, on identical standards to avoid confusion. Furthermore, the costs of protective actions may vary from one region to another so that a universal response set would lead to suboptimal solutions. Therefore, responses have to be more flexible, but should be predetermined for each local area. The objective of the flexible response strategy is to accomplish an identical level of public protection using different means.

(4) In addition to public actions based on collective decision making, all citizens should be given the opportunity to increase voluntarily their desired level of safety. This requirement implies, first, that all measures are publicized in the local media and, second, that recommendations about additional means of self-protection are communicated and explained. Leaving actions entirely to individuals would probably lead to an unacceptable violation of equity, because less educated persons find it difficult to make voluntary adjustments and would, therefore, face a much higher risk. Combining an adequate collective safety standard with the opportunity to further reduce risk at an individual level appears to represent the best trade-off between equity and freedom of protective action.

(5) Confidence in the above measures depends on the capability of the emergency manager to put the risk and the effectiveness of countermeasures in perspective, i.e. in relation to other risk situations. Risk comparisons are not well received by most members of the public, because they have been extensively used to justify nuclear power or other low probability/high-consequence risks. Abstract risk figures, however, have hardly any meaning for most of the public. Furthermore, the way in which risk figures are presented (for example, in percentages of additional cancers or in absolute numbers of additional cancers) makes a strong difference to their public perception. Hence, it seems advisable to use the abstract figures in combination with one or two other related risks. The most acceptable approach is to use a reference risk that is also technological in nature and involuntary, such as the risk from food additives or air pollutants. Publishing those risk figures prior to any emergency would be helpful in preparing the public to deal with probabilities and in soliciting responses by interest groups.

(6) For the purpose of gaining public confidence, an institutional separation of emergency response activities and nuclear energy licensing or even promotion is essential. Public concern in Italy peaked when it became known that ENEA (Ente Nazionale Energie Nucleare e Alternative) was responsible for both the licensing of

nuclear power plants and the control of public safety (Otway *et al.*, 1987). Public recommendations were followed more often in those countries where the government was not perceived as an involved party in the nuclear debate.

(7) Apart from any nuclear emergency, the handling of modern technological hazards requires a better understanding of the meaning of probabilities and the risk management process. In the long run, educational programs for schools and professional training should be introduced so that probabilistic thinking is slowly incorporated into the generally accepted notion of common sense. Deterministic heuristics, which still predominate common sense reasoning, prevent many people from evaluating risks consistently and responding to emergencies in a rational manner.

Even if all these suggestions were implemented, over-reaction on the one hand and apathy on the other will still be likely responses to nuclear crises. But a more consistent approach to risk communication and a better preparation for nuclear emergencies could certainly increase the proportion of adequate responses and enhance public protection. Some countries, such as West Germany, have already started to reform their emergency response system. Although the intended unification of intervention thresholds is going in the right direction, the drive to centralize the response system may result in inflexibility and inability to cope with high local variations in exposure. A viable compromise between centralized guidelines and flexible reactions based on local conditions is probably the best solution.

Conclusions

The accident of Chernobyl did not only leave its mark in the form of radioactive fallout in most European countries, but also had a lasting impact on public opinion and personal attitudes. Attitude theory suggests that in countries with less visible impacts and low percentages of uncommitted or 'don't know' responses, attitude changes will be merely temporary. In most countries affected by the fallout, citizens responded with a more negative attitude toward nuclear power in general, but had difficulties in finding appropriate protective action. A small, but verbal minority took major protective actions which most experts estimated as 'overdone', while the majority did not even comply with rather simple government recommendations.

Antinuclear sentiments among the public peaked in the immediate aftermath of the accident and shifted from the perceived threat of the fallout to the evaluation of the domestic nuclear program. During the first year after the accident, however, most former adherents of nuclear energy were again supporting the use of nuclear power and opted for a moderate share of nuclear energy in their native countries. This is particularly true for countries with a strong nuclear program. People there find enough positive cues and feel comforted by the perception of a majority in favor of using nuclear power. It is known that the perception of 'swimming along with the majority' has a strong impact on attitude recovery and attitude persistence (Chaiken & Stangor, 1987). However, even those in favor of the domestic use of nuclear power expect changes in domestic policies, in particular in redesigning emergency response systems. Although governmental management institutions were still respected and perceived as at least partially credible, competing information from antinuclear groups were equally well-perceived and trusted.

In many instances, Chernobyl forced governments to respond immediately to public pressure and to reconsider or alter their existing nuclear policies. The most striking political reaction in most countries was not to abolish nuclear power, but to reconsider and possibly delay the further construction of nuclear facilities. Countries with commitments to phase out or abolish nuclear energy, such as Austria, Denmark, Ireland, and Sweden, reconfirmed their nation's decision, whilst most countries with ongoing nuclear programs continued to support the further use of nuclear energy but at a lower growth rate or with a longer implementation time. However, many opposition parties in countries with ongoing nuclear programs changed their nuclear policy and pursued their antinuclear position more radically.

The overall analysis reveals that the primary incentive for European governments in responding to the crisis was to initiate dose reduction measures in accordance with the amount of radionuclides dispersed in each country. In addition, the more outraged the public reacted to the accident, the more public officials felt obliged to increase or intensify the measures for dose reductions. Both public outrage and dose savings were less severe in countries with a high share of nuclear power, thus suggesting that public pressure was the primary factor for the release of dose reduction measures, while the share of nuclear energy was only indirectly related to dose savings.

The official stance on nuclear energy has been and continues to be defensive and under severe scrutiny, but the case is certainly not lost for those governments in favor of nuclear energy. In Europe as well as in the United States, Chernobyl triggered the attention of risk manager and the public alike to the question of post-accidental exposure and its management. The growing interest in the ability of official risk managers to plan and implement adequate protective actions has already influenced the licensing procedure for Seabrook (New Hampshire, U.S.A.) and was one of the reasons for abandoning the plan to build a reprocessing plant in Wackersdorf (Bavaria, West Germany). Many public groups demand that risk management should be better prepared to handle emergency situations and that existing plans for evacuation or food monitoring should be re-examined and tested.

After Chernobyl, it has also become more difficult to confine emergency planning to design-based accidents. New nuclear facilities will probably not be licensed unless enough evidence is provided to show that protective actions are sufficient and management provisions adequate to ensure public protection even for a severe accident beyond the design criteria. If the nuclear community is flexible enough to meet this challenge and to introduce new convincing concepts of providing emergency responses for Chernobyl-type accidents including changes in reactor design which physically exclude the possibility of large accidents, the initial shock of Chernobyl and its reflections on public attitude may be overcome and a new era of nuclear power may arise. Otherwise, the fate of nuclear power will be doubtful; the inoculation effect of attitude persistence will probably be insufficient to immunize people from a second Chernobyl should it ever occur.

References

Allensbach, Institut für Demoskopie. (1987). *Public Opinion Poll on Nuclear Energy After Chernobyl.* Report for the Atomic Industrial Forum. FRG: Allensbach.

Belelli, U. (1988). Public and media attitudes to nuclear power in Italy. In Uranium Institut. Ed., *Uranium and Nuclear Energy: 1987. Proceedings of the Twelfth International Symposium,* London, September 204. London: Butterworth.

Chaiken, S. (1980). Heuristic versus systematic information processing and the use of source versus message cues in persuasion. *Pers. Social Psychology* **39**, 752–766.

Chaiken, S. & Stangor, C. (1987). Attitudes and attitude change. *Annual Review of Psychology* **38**, 575–630.

Cotton, J. L. (1985). Cognitive dissonance in selective exposure. In D. Zillman & J. Bryant, Eds., *Selective Exposure to Communication* (pp. 11–33). Hillsdale, NJ: Erlbaum.

Covello, V. T., Von Winterfeldt, D. & Slovic, P. (1986). Risk communication: A review of the literature. *Risk Abstracts* **3**, 171–182.

Department of Energy. (1987). *Health and Environmental Consequences of the Chernobyl Nuclear Power Accident.* DOE/ER-0332. Springfield, VA: National Technical Information Service.

European Community (1984). *Change of directives to assure the protection of the population and therefore against the dangers of ionizing radiation.* Bulletin of the Europeam Community 265(27) (October 5). 84/467/EURATOM.

Festinger, L. (1957). *A Theory of Cognitive Dissonance.* Stanford, CA: Stanford University Press.

Flavin, C. (1987). Chernobyl: The political fallout in western Europe. *Forum for Applied Research and Public Policy (Summer)*: 16–28.

Frey, D. (1986). Recent research on selective exposure to information. *Advances in Experimental Social Psychology*, **19**, 41–80.

Gesetz zum vorsorgenden Schutz der Bevölkerung gegen Strahlenbelastung. 1986. (Strahlen-schutzvorsorgegesetz-StrVG) *Bundesgesetzblatt*, **1**, 2610–2614.

Hohenemser, C., Deicher, M., Ernst, A., Hofsass, H., Lindner, G. & Rechnagel, E. (1986). Chernobyl: an early report. *Environment* **28**, 6–13; 30–43.

Hohenemser, C. & Renn, O. (1988). Shifting public perceptions of nuclear risk: Chernobyl's other legacy. *Environment*, **30**, 4–11; 40–45.

Joutsenniemi, A. (1987). *Chernobyl: Information and media coverage in Finland.* Department of Communication at the University of Helsinki and the Research Institute for Social Sciences at the University of Tampere. Helsinki (Tampere, Finland) August 25, 1987.

McGuire, W. J. (1985). Attitude and attitude change. In G. Lindzey & E. Aronson, Eds., *Handbook of Social Psychology*, **3**, 223–346. New York, NY: Random House.

Midden, C. J. H. & Verplanken, B. (1990). The stability of nuclear attitudes after Chernobyl. *Journal of Environmental Psychology*, **10**, 111–119.

Newsweek (1986). U.S. fears and doubts: A Newsweek Poll. **30**, *Nucleonic Week* (1986). Antinuclear fallout from Chernobyl continues to wash over Europe. *May*, 11–13.

Nuclear Energy Agency (NEA). (1988). *The Radiological Impact of the Chernobyl Accident in OECD Countries.* Paris: OECD-NEA.

Otway, H., Haastrup, P., Cannell, W., Gianitsopoulos, G. & Paruccini, M. (1987). *An Analysis of the Print Media in Europe Following the Chernobyl Accident.* Report of the Joint Research Center of the Commission of the European Community. Ispra, Italy: EC.

Peters, H. P., Albrecht, G., Hennen, L. & Stegelmann, H. U. (1987). *Die Reaktionen der Bevölkerung auf die Ereignisse in Tschernobyl. Ergebnisse einer Befragung.* Report of the Nuclear Research Center, Jülich, Jül-'Spez-400. Jülich, F.R.G.: KFA.

Peters, H. P., Albrecht, G., Hennen, L. & Stegelmann, H. U. (1990). 'Chernobyl' and the nuclear power issue in West German public opinion. *Journal of Environmental Psychology*, **10**, 121–134.

Petty, R. E. & Cacioppo, E. (1986). The elaboration likelihood model of persuasion. *Advances in Experimental Social Psychology*, **19**, 123–205.

Rager, G., Klaus, E. & Thyen, E. (1987). Der Reaktorunfall in Tschernobyl und seine Folgen in den Medien. Eine inhaltsanalytische Untersuchung. Manuscript, Lecture at the University of Karlsruhe (Karlsruhe, FRG, June 3, 1987).

Reisch, F. (1987). The Chernobyl accident—its impact on Sweden. *Nuclear Safety*, **28**, 29–36.

Renn, O. (1988). Public responses to Chernobyl: Lessons for risk management and communication. In Uraniun Institut Ed., *Uraniun and Nuclear Energy: 1987. Proceedings of the Twelfth International Symposium*, London, Sept. 2–4, 1987, pp. 53–66. London: Butterworth.

Renn, O. (1986). *Gedanken und Reflexionen nach dem Unfall von Tschernobyl.* Report of the Nuclear Research Center of Jülich. Jülich, FRG: KFA.

Renn, O. (1984). *Risikowahrnehmung der Kernenergie.* Frankfurt and New York: Campus.

Roser, T. (1988). The social and political impact of Chernobyl in the Federal Republic of Germany. In Uraniun Institut Ed., *Uraniun and Nuclear Energy: 1987. Proceedings of the Twelfth International Symposium*, London, Sept. 2–4, 1987. London: Butterworth.

Strahlenschutzkommission. (1987). *Scientific Basic Concepts for the Assessment of Derived Levels of Dose and Contamination According to Section 6 of the Precautionary Radiological Protection Act.* Strahlenschutzvorsorgegesetz. Bonn, FRG: SSK.

Suhonen, P. & Virtanen, H. (1987). *Public Reaction to the Chernobyl Nuclear Accident.* Tampere, Finland: Research Institute for Social Sciences, University of Tampere.

Wall Street Journal. (1986). The NBC news poll on Chernobyl, May 2, **8**.

Washington Post. (1986). Reactions to Chernobyl in Europe, May 5, **13**.

Wallmann, W. (1987). Chernobyl's impact in West Germany. *Forum for Applied Research and Public Policy*, Summer, 13–15.

World Energy Conference. (1989). Attitudes toward energy systems. A multinational comparison. In WEC Committee Ed., *Energy and the Public.* London: WEC.

WHAT WAS THE MEANING OF CHERNOBYL?*

TIMOTHY C. EARLE and GEORGE CVETKOVICH

Western Institute for Social and Organizational Research, Psychology Department, Western Washington University, Bellingham, Washington, U.S.A.

Introduction

The Chernobyl nuclear power plant accident of 1986 appears to have meant many things for many people. Chernobyl was new information. For some, it was information about a particular power plant and its management. For others, however, it may also have been information about nuclear power in general or even about hazardous technologies world wide. Chernobyl was attended to, found to have meaning, judged to be personally relevant and even acted upon by some people. Others found it totally lacking in interest. This diversity of reactions to Chernobyl and the reasons for it are the underlying themes of this special issue. Like their subjects, the contributors to this issue were variously motivated. In reading each of these articles, we attempted to answer this question from the authors' points of view. Apart from personal curiosity about the effects of a major supranational technological disaster, why should we be interested in public and governmental reactions to the Chernobyl accident? Some authors (Eiser *et al.*, 1990; Midden & Verplanken, 1990) appear to have seen Chernobyl primarily as an opportunity to explore theoretical and methodological issues within environmental psychology. These researchers, along with all the others, also used the Chernobyl accident to study the effects of a nuclear power plant disaster on public attitudes and behaviors. A third group (Drottz & Sjoberg, 1990; Peters *et al.*, 1990; Renn, 1990) explored the social implications of the Chernobyl accident, particularly with regard to communicating risks and managing hazards. As a result of these studies, our understanding in each of these areas, from rather narrow social science issues to broad societal problems, has been advanced.

Our comments on this collection are not comprehensive: van der Pligt and Midden (1990) have provided a useful review of the main issues and results in their introduction. We have focused instead on three general areas of interest. Firstly, we have tried to identify issues that we believe are or should be controversial and in need of closer examination. Secondly we have tried to point out blank areas and new directions for research. Thirdly, we have tried to discover an underlying meaning to Chernobyl that is supported by all of the contributions to this special issue.

Theoretical and methodological issues
Several of these studies of public thinking and behavior are notable for the use of effective methods such as cross-national and cross-time comparisons. Two issues, the concept of defensive avoidance and within-subject analyses, deserve specific comment.

* This material is based upon work supported by the National Science Foundation under Grant No. SES-8822356.

Defensive avoidance. In our view, the concept of decision-making styles, which is the focus of the cross-national study by Eiser *et al.* (1990), is less controversial than it should be. There seems, for example, to be a general and rather uncritical acceptance by many researchers of 'defensive avoidance' as an important and widely used public decision-making strategy. The depth of this acceptance can be seen also in the Drottz and Sjoberg article (1990) where confident reference is made to the 'denial of risks'. Our objection is simple: The concepts of defensive avoidance and denial are based on the assumption of a correct, objective viewpoint. But risk judgement is fundamentally subjective (Fischhoff *et al.*, 1984).

In analysing public responses to Chernobyl, Eiser *et al.*, argue that persons with low involvement (interest and concern) were engaging in defensive avoidance. But the measures for both these concepts can be interpreted as simple lack of interest. In another context, consider the individual who states that he is unconcerned about the negative health effects of the drugs he is using. Is he 'denying' those effects, or is he genuinely uninterested? This is a question that is of fundamental importance to risk communication and hazard management. If we accept defensive avoidance and denial (as many hazard managers have), then a strategy based on education to the facts (the correct viewpoint) is implied. This strategy, however, is criticized by Eiser *et al.*, and others on the basis that persons with different attitudes will judge the salience of the facts differently. If, on the other hand, we take the individual at his word and accept his indifference to what interests us, then a strategy based on learning what *is* important to the individual is implied.

In questioning the use of the concepts of defensive avoidance and denial, we are not claiming that all applications are invalid. We suggest instead that it is an empirical question that can be addressed, for example by process data, collected as individuals make decisions in supposed conflictual contexts. Knowledge of how people actually think when they are making decisions (whether or not they are engaging in denial or any other strategy) should improve our ability to communicate about and manage hazards.

Within-subjects analyses. In their careful study of the stability of nuclear power attitudes after Chernobyl, Midden and Verplanken (1990) demonstrate the importance of studying individuals if we wish to understand psychological processes. Data at the aggregate level can hide significant variations at the individual level. Although generally ignored in the past, it is becoming clear that knowledge of individual psychological processes is crucial to effective risk communication. This is so because risk communicators must deal with significant individual and group differences. These differences in audience interests, values, knowledge, etc. can form the basis for the design of effective communication strategies (Earle *et al.*, 1989).

Midden and Verplanken (1990) show not only that supporters and opponents of nuclear power based their judgements on different dimensions (predominant benefits vs predominant risks) but that the former were more ambivalent and less stable than the latter. Proponents of nuclear power were more vulnerable to the negative news from Chernobyl. It is of course unknown how opponents of nuclear power would react to a positive news event that was of the magnitude of Chernobyl.

Effects on public attitudes and behaviors
A central strength of this collection of studies is the large number of interesting findings regarding public attitudes and behavior, and the relationship between the two. Some

results, such as Drottz and Sjöberg's (1990) findings that individuals were more optimistic about risks to themselves than risks to others, have solid foundations in earlier studies (e.g. Weinstein, 1987). Others, such as those dealing with cross-national comparisons, broke new ground. We look at some of each.

Individual differences. Drottz and Sjöberg (1990) describe several intriguing contrasts among the groups they studied in Sweden. One of these is between adolescents and adults. In general, adolescents were less worried by Chernobyl and changed their behavior less in response than did adults. Drottz and Sjöberg point out that little is known about adolescents' views on the risks of nuclear power. To our knowledge this is true, but a good deal is known about adolescents' views on other hazards such as the use of alcohol, tobacco and automobiles (US Bureau of Alcohol, Tobacco and Firearms, 1988; Jessor, 1987; Urberg & Robbins, undated). In each of these cases, adolescents see their personal risks as being low relative to those of older persons. These age differences have been attributed to the tendencies of young people to focus on the present and on themselves and also to their overconfidence in their ability to control risks.

Drottz and Sjöberg also describe women's reactions to Chernobyl as being more negative than those of men, and those of locally affected farmers as being greater than those of the general public. These authors and others also show that the public in general has a more negative view of nuclear power than the experts of that technology. In the past, these and other differences between groups have made communication about nuclear power difficult. That was because the differences were either ignored or interpreted as unfortunate deviations from the correct point of view and susceptible to remedial education. Today, these differences tend to be interpreted as legitimate expressions of social and cultural diversity. Knowledge of these differences could therefore contribute to the development of effective communication strategies by identifying areas of concern and common interest.

Informative as it is about group differences, it seems unlikely, however, that survey research, such as that of Drottz and Sjöberg, can contribute a great deal toward this effort (nor is it designed to). This is because much of this work assumes a certain cognitive structure with relationships between attitudes, beliefs and values. Survey results, as in the case of Drottz and Sjöberg, show that attitudes toward nuclear power can be predicted by other attitudes, values and beliefs. But respondents' cognitive representations of nuclear power may be unrelated to these elements and thus unaffected by information about them. This view is consistent with contemporary models of human information processing (Wyer & Srull, 1986). A more promising approach would begin with the exploration of how information about Chernobyl (or any other hazard) is actually represented cognitively by people. This might take the form of a process study, such as Pennington and Hastie's (1986) investigation of how jury members represent trial evidence. The point is that we should do whatever possible by whichever means, to uncover the diversity of hazard representation among the public. Effective risk communication and hazard management may depend on it.

Inadequate information and uncertain reactions. The finding by Peters *et al.*, that uncertainty and insecurity were the dominant public reactions to Chernobyl in West Germany, demonstrates that the effects of a risk message on an audience are the product of more than just the personal relevance and meaning of the new information. An effective message is also easily understood and can offer useful guidance on what to do. Chernobyl was a new hazard to most people; the type of event about which the public knew little, in particular as regards radiation. As a consequence, the public was dependent upon expert advice communicated via the news media. Although the news media did a good job (a point we will elaborate on later), they had a very difficult task in trying to produce clear, useful messages.

Like policy makers, the news media had to make practical sense out of often contradictory and uncertain technical information. As Hammond *et al.* (1983) have argued, these and other factors form fundamental obstacles to the use of scientific information in the making of public policy and (we would add) personal protective action decisions. The news media had also to communicate useful information on a hazard whose health effects may not become apparent for many years. Such threats of delayed health effects from radiation and other technological hazards can lead to significant stress (Baum *et al.*, 1983). One can attempt to limit exposure to radiation, but, as Renn points out in this issue, once exposed there is little one can do about it except overreact and/or worry, and this may lead to illness. From the public point of view, the information in the Chernobyl risk message was, in general, poor. For most people, the low quality of the new information combined with a lack of knowledge about radiation produced a low capacity to act. People did not know what to do. They therefore decided to await the arrival of newer, more useful information before acting. Future research in this area should be aimed at the accurate assessment of audience information needs and the design of messages to meet them.

The spread of credibility. Effective risk messages must offer guidance, be clear and also *credible*. Peters *et al.* (1990) found that in West Germany a wide variety of sources were judged by the public to produce credible messages about Chernobyl. The 'official', expert word was not necessarily the last word. This result illustrates what appears to be a growing, general trend in industrialized countries, namely away from management by small elites and towards management based on the views of diverse social groups. This can be seen generally in the key role played by stakeholder groups in hazard management (Mazur, 1981). With respect to Chernobyl, the trend was seen, for example, in the development of 'informal channels of communication' in Poland (Wyka & Glinski, 1989). The implications for risk communication of this 'spread of credibility' from a single official source to multiple sources are outlined later in this paper.

Attitude change across nations. As part of his excellent review of several aspects of the Chernobyl case, Renn offers a very interesting and useful analysis of the relative duration of attitude change across countries. Renn points out that the level of concern within a country was proportional to the 'real radiation danger' (whole body dose). This indicates that Chernobyl can be interpreted as an example of successful risk communication. Yet success was limited. As we discussed above, the information in the Chernobyl message was insufficient to enable people to take appropriate protective action. It was effective, however, in raising public conern. Inaction (or over-reaction relative to the expert view) was due to low capacity to act and not to a low involvement or lack of interest.

Renn also points out that the countries with the most dramatic attitude changes were those with high levels of radiation exposure and low initial levels of involvement or interest in nuclear issues. The low initial levels of involvement meant that these people had no easily accessible concepts to understand nuclear power. Their subsequent personal experience with the fallout from Chernobyl (in addition to the great volume of information available through other persons and the news media) established new, vivid concepts in their memories producing a negative view of nuclear power. As Weinstein has documented in his recent comprehensive review of the powerful effects of personal experience on self-protective behavior (Weinstein, 1989), these concepts will guide their future attitudes and behavior.

With regard to the general problem of cross-national comparisons, it seems that, despite some intriguing data (e.g. Eiser *et al.*, 1990), little is made of them in this collection of papers. The relative lack of prior research in this area indicates that cross-national comparisons are not easy to tie down. The recent work of Bastide *et al.* (1989), suggests however that the job is not impossible and that the results can be both interesting and useful.

Risk communication and hazard management
It is no doubt the hope of many that we can learn from Chernobyl to avoid similar events in the future. The lessons of Chernobyl for hazard management were discussed by several authors. Hazards can be managed in three general ways: (a) through improved engineering and technology; (b) through regulation; and (c) through communication. Since all of the authors represented here are social scientists, the last of these management methods received most attention. It is important to keep in mind, however, that risk communication is only one of several approaches to hazard management, and, like the others, is limited.

The role of the news media. Obviously the news media play an important role in risk communication and hazard management, without the news media we would be dependent for hazard information on our own personal experience or on whatever we learn directly from other people. Nevertheless, just as everyone agrees that the news media are important, everyone is a media critic. These attacks generally stress two basic points: (a) That the news media focus on events and consequences rather than on the hazards that caused them; and (b) The news media pay too much attention to hazards with risks having very high consequences but very low probabilities. Chernobyl therefore was a prime target for both the news media and their critics. In discussing the Chernobyl case, Renn provides a needed antidote to what he calls 'the common prejudice' that the news media are somehow responsible for public reactions to hazards. As argued elsewhere (Earle & Lindell, 1980), and as Renn demonstrates with regard to Chernobyl, the news media tend to accurately reflect the information needs of their audiences. When threatened by a hazard, the public has little need for technical detail; what is needed is information on how to cope, not information about the hazard but information about hazard management. And the news media generally do an excellent job of providing this vital information.

Self concept and risk communication. In the conclusion to their analysis of the effects of Chernobyl on the Swedish population, Drottz and Sjöberg hint at a key factor in determining the success or failure of risk communication. These authors suggest effective risk communication must be based on 'the crucial aspects of information' that determine the audience's risk attitudes (A similar point is made by Eiser *et al.*). This seems clear. One group of people, for example, may be interested in a (hazardous) technology because of the benefits it provides, while another group may be interested in the same technology because of its costs. The former group will be motivated to process information about benefits but not costs; the motivation of the latter group will be just the opposite. To be successful, a communicator must be able to tailor the message to the group. As Drottz and Sjöberg indicate, the features of a hazard in which an individual is interested are determined by his/her values and goals. These are contained in one's self concept. Unfortunately for risk communicators, an individual can access a variety of self concepts that vary with context (Wyer & Srull, 1986). Thus, in one situation a person may access a concept in which he/she highly values personal health. In another context, the same person may access a concept in which personal pleasure is the highest value. This basic lability of self concept is one of several factors that make effective risk communication a rather more demanding task than it may at first appear to be.

Public acceptance of normal accidents. According to Peters *et al.*, the Chernobyl message effected few political consequences in West Germany. In their explanation, these authors appeal to the concept of 'normal accidents' (Perrow, 1984) which they interpret as 'the price one has to pay for the highly industrialized society' in West Germany. That is, West Germans accept this price as part of the management of the nuclear hazard. We question this explanation on two grounds. Firstly, there is a number of plausible alternative explanations for the lack of political effects: (a) there were strong prior beliefs, particularly against nuclear power, and therefore little room for movement; (b) there was a lack of effective actions to be taken; and (c) West Germans believed that their nuclear industry was better managed than the Russians'. Secondly, 'normal accidents' are not the 'accepted price' of industrialization. According to Perrow, normal accidents are the inevitable result of multiple and unexpected failures in systems with two special characteristics, 'interactive complexity' and 'tight coupling'. Nuclear power plants are good examples of this type of system. Normal accidents are rare, but some can result in catastrophes. Acceptance of these accidents would therefore be a difficult management strategy to sell, well beyond the limits of risk communication. A technical/engineering solution such as that suggested by Renn (a new reactor design that physically excludes the possibility of catastrophic accidents) would seem more feasible.

Planning for risk communication and hazard management. The hazard management lessons of Chernobyl spelled out by Renn (1990) are similar in some respects to the lessons learned from Three Mile Island. Renn breaks new ground, however, when he proposes improvements in public participation and risk communication. Although Renn attempts to step away from an expert/technocratic view of risk communication, he does not go far enough. This can be seen in his depiction of the public as independent decision makers, similar in many ways to the experts. In this context, risk communication is simply a problem of training people to process technical information and of supplying it to them when needed. In our view, this concept of risk communication is idealistic. Realistically, most people do not want, do not need and are unable to use technical information about the hazard. People primarily want information about hazard management. They want to know what to do and what is being done by others. This view may appear to underestimate the public, but effective risk communication and hazard management must be based on realistic assessments of the interests, goals and cognitive capacities of the audience, not on idealized conceptions.

Conclusions

In reviewing this collection of studies on reactions to the Chernobyl nuclear power plant accident, we have focused on three general areas of interest. First, we have pointed to some controversial issues. For example the use of concepts such as defensive avoidance and denial. These concepts seem to impose problem definitions from the outside that may clash with legitimate subjective points of view. Another controversial issue was the utility of survey research in understanding how people think about hazards. Use of survey research often seems to involve the imposition of problem structures on respondents. This is often appropriate and produces reliable results (Fischhoff *et al.*, 1987). But effective risk communication and hazard management depend on knowledge of how respondents structure problems under natural conditions. For these purposes, use of techniques less intrusive than traditional survey research may be necessary.

Second, we tried to identify productive directions for new research. Following on from our comments above, we believe knowledge of how people actually think about hazards in natural contexts is needed. As the studies collected in this special issue demonstrate, people react to hazardous events in a wide variety of ways. In the light of this, social science researchers should work, firstly, to understand these diverse social and cultural reactions better and, secondly, to develop methods of community decision making that can generate effective hazard management policies based on these diverse views.

Third, we tried to uncover an underlying or general meaning of Chernobyl. All of these studies support the conclusion that social knowledge should form the basis for hazard management policy. The notion that information about people can be as valuable in the management of hazards as information about technologies is relatively new and not universally accepted. But reduced public confidence in technical knowledge, particularly with regard to nuclear power, has shifted the main management burden to social knowledge. In this process, the accepted general model of risk communication has evolved from 'technocentric' (*from* experts *to* laypersons) to 'sociocentric' (*among* all interested parties). In our view, the underlying meaning of Chernobyl, and its greatest impact, is its contribution to this general trend. Chernobyl has led to the further undermining of management based on technocratic elites. Chernobyl thus presents a challenge to social scientists to help people to learn more about one another and to learn to work together to solve their problems.

References

Bastide, S., Moatti, J-P., Pages, J-P. & Fagnani, F. (1989). Risk perception and social acceptability of technologies: The French case. *Risk Analysis*, **9**, 215–223.

Baum, A., Fleming, R. & Davidson, L. M. (1983). Natural disaster and technological catastrophe. *Environment and Behavior*, **15**, 333–354.

Drottz, B-M. & Sjöberg, L. (1990). Risk perception and worries after the Chernobyl accident. *Journal of Environmental Psychology*, **10**, 135–149.

Earle, T. C. & Lindell, M. K. (1980). *The Role of the News Media in the Gasoline Crisis*, BHARC-411/80/02. Seattle: Battelle Human Affairs Research Centers.

Earle, T. C., Cvetkovich, G. & Slovic, P. (1989). *The Effects of Involvement, Relevance and Ability on Risk Communication Effectiveness*. Presented at the Subjective Probability, Utility and Decision Making (SPUDM) Meeting in Moscow, USSR.

Eiser, J. R., Hannover, B., Mann, L., Morin, M., Van der Pligt, J. & Webley, P. (1980). Nuclear attitudes after Chernobyl: a cross-national study. *Journal of Environmental Psychology*, **10**, 101–110.

Fischhoff, B., Watson, S. & Hope, C. (1984). Defining risk. *Policy Analysis*, **17**, 123–139.

Fischhoff, B., Svenson, O. & Slovic, P. (1987). Active responses to environmental hazards: Perceptions and decision making. In D. Stokols & I. Altman, Eds., *Handbook of Environmental Psychology*. New York: Wiley.

Hammond, K. R., Mumpower, J., Dennis, R. L., Fitch, S. & Crumpacker, W. (1983). Fundamental obstacles to the use of scientific information in public policy making. *Technological Forecasting and Social Change*, 24, 287–297.

Jessor, R. (1987). Risky behavior and adolescent problem behavior: An extension of problem behavior theory. *Alcohol, Drugs and Driving*, 3, 1–11.

Mazur, A. (1981). *The Dynamics of Technical Controversy*. Washington, D.C.: Communication Press.

Midden, C. J. H. & Verplanken, B. (1990). The stability of nuclear attitudes after Chernobyl. *Journal of Environmental Psychology*, 10, 111–119.

Pennington, N. & Hastie, R. (1986). Evidence evaluation in complex decision making. *Journal of Personality and Social Psychology*, 51, 242–258.

Perrow, C. (1984). *Normal Accidents: Living with High-Risk Technologies*. New York: Basic Books.

Peters, H. P., Albrecht, G., Hennen, L., Stegelmann, H. U. (1990). 'Chernobyl' and the nuclear power issue in West German public opinion. *Journal of Environmental Psychology*, 10, 121–134.

Renn, O. (1990). Public responses to the Chernobyl accident. *Journal of Environmental Psychology*, 10, 151–167.

Urberg, K. & Robbins, R. (undated). Adolescent invulnerability: Developmental course and relationship to a risk-taking behavior. Unpublished manuscript. Department of Family and Consumer Resources, Wayne State University. Detroit, Michigan.

U.S. Bureau of Alcohol, Tobacco and Firearms. (1988). *Final Report, Research Study of Public Opinion Concerning Warning Labels on Containers of Alcohol Beverages*. Vol. II.

Van der Pligt, J. & Midden, C. J. H. (1990). Chernobyl: three years later. *Journal of Environmental Psychology*, 10, 91–99.

Weinstein, N. D. (1987). Unrealistic optimism about illness susceptibility: Conclusions from a community-wide sample. *Journal of Behavioral Medicine*, 10, 481–500.

Weinstein, N. D. (1989). Effects of personal experience on self-protective behavior. *Psychological Bulletin*, 105, 31–50.

Wyer, R. S. & Srull, T. K. (1986). Human cognition in its social context. *Psychological Review*, 93, 322–359.

Wyka, A. & Glinski, P. (1989). *Polish Public Reaction to Nuclear and Fossil Energy*. Paper presented at the Workshop on Energy and the Environment, Paris, October 26–27.

PUBLIC PERCEPTIONS OF ELECTRIC POWER TRANSMISSION LINES

LITA FURBY,* PAUL SLOVIC,** BARUCH FISCHHOFF and ROBIN GREGORY**

Eugene Research Institute, 474 Willamette Street, Suite 310, Eugene, Oregon 97401, U.S.A.

Abstract

Electric power transmission lines have recently met a very significant amount of public opposition. The source of such opposition varies from case to case, and is often hard to identify. Stated objections have included land use conflicts, noise created by the lines, aesthetic concerns, and fears of health and safety threats. Despite the sometimes enormous costs and long delays caused by strong opposition to transmission line siting and construction, both utilities and governmental regulators seem baffled at why the public objects so vehemently. At the same time, opponents are often equally baffled at why their objections seem to go unheeded. As a step toward developing satisfactory solutions to the conflict, this article reviews and critiques the literature dealing with attitudes toward electric power transmission lines, and outlines a conceptual framework for understanding the determinants of those attitudes.

The purpose of this article is to review and critique the literature dealing with the formation and expression of attitudes related to the siting and construction of electric power transmission lines, and to develop a conceptual framework that outlines the determinants of those attitudes. Our goal is to provide a useful tool both for understanding public reactions to powerlines in the past, and for improving the powerline siting and construction process in the future.

General Background

Current and future transmission line needs

High-voltage 345 kV and above) electric transmission lines are an increasingly important part of our electrical supply system. Growing demand for electricity and difficulties of developing new electricity sources have augmented the need for transmission capability, which is achieved most efficiently by transmission at very high voltages. One complicating factor in planning such lines is the unpredictability of demand. In particular, recent demand projections have often been too high, leaving utilities open to charges of exaggerating the need for electricity in order to promote their own growth and justify new power generation at the ratepayers' expense. On the other hand, the utilities would also have been open to criticism if there had ever been a shortage of electricity: Americans are unaccustomed to and generally intolerant of interruptions in their power supply. Given the increasing costs and difficulties of siting and constructing new generating sources, one attractive method of assuring sufficient and reliable power is for utilities to share sources.

---

* To whom request for reprints should be addressed.
** Decision Research, 1201 Oak Street, Eugene, Oregon, 97401, U.S.A.

Flexible intraregional and interregional transfer arrangements allow one geographical area to alleviate temporary shortages in another area. Permanent arrangements allow regions with an abundance of power (e.g., the Northwest with its hydropower) to provide electricity to regions with less generating capacity (e.g., Southern California).

Significance of public attitudes toward transmission lines
Transmission lines currently represent a problem area in the electric power system: they require considerable land for their corridors, and the use of that land for transmission lines may conflict with other land use practices or plans; they cause noise; they are perceived as visually unattractive; and they are perceived to cause health problems and safety risks for both animals and humans. As a result, high-voltage transmission lines have recently met a very significant amount of public opposition. Severe conflict was apparent during the 1960s in such states as Ohio, Virginia, Oregon and California (Fricke, 1964; Young, 1973; Mason, 1982). The 1970s saw even more intense opposition with Minnesota (Casper and Wellstone, 1981) and New York (Ray, 1978) being perhaps the most publicized cases, but others included those in Montana and Washington (Mason, 1982), South Dakota (Yankton, 1981), Ontario (Boyer *et al.*, 1978), Arizona and California (Mason, 1982) and Texas (Young, 1978). Opposition to transmission line siting and construction has sometimes caused enormous costs to the utilities, through long delays in gaining regulatory approval, litigation fees, and occasionally even vandalism. Such confrontations suggest that something about the decision-making process regarding siting and construction may not adequately address the public's concerns. In order to develop satisfactory solutions to this conflict, it is important to understand public attitudes toward transmission lines better.

Conceptual Framework

Overview
In our review of the literature we identified a set of key elements that appear to be important determinants of attitudes related to the siting and construction of transmission lines. Figure 1 presents these elements and a suggested scheme for conceptualizing the role each plays in determining the attitude formation process. Figure 1 is meant to be more of an organizing schema than a model of attitude formation and behavior. We are far from the level of understanding needed for precise modeling.

Some of these elements are fairly direct effects of the lines: property value effects, aesthetics, human health and safety effects, environmental effects, and economic benefits. Others are largely management-related issues: equity effects, process characteristics, and (to some extent) information and knowledge. For elements in each of these categories, people's perception of the element, along with their values associated with it, can play a role in determining their attitudes toward a powerline. Still other elements do not fall into either of these categories but instead overlay the entire framework: symbolic meaning, amount of information and knowledge, and conflict dynamics.

In the following sections, we review the literature pertaining to each of these key elements. Following the literature review, we discuss the most profitable directions of future research.

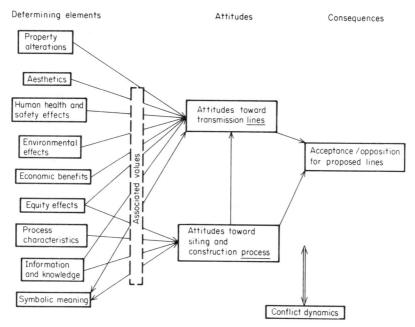

FIGURE 1. Conceptual framework.

Historical perspective

Attitudes toward powerline rights-of-way and easements, particularly in rural areas, can be fully understood only if we examine the economics involved from an historical perspective.

> The years from 1920 to 1930, when the rights of way for many of our cross-country transmission lines were purchased, was a period of low land values and depressed grain markets. Many farmers were in debt and had heavy mortgages. They were glad when an opportunity arose to sell rights of way for cash and gave little thought to the depreciation or severance damage resulting from such easements. Today [1955] the picture is entirely different and one in which we can hardly recognize the farmer we saw in the twenties. He is much more of a businessman and certainly under much less financial pressure. He has been subsidized by the government with high prices assured for his grain. After years of prosperity he is no longer in debt. Land values have increased, more intense use is made of the land, and he no longer wants the power line on his property (Crawford, 1955, p. 37).

Consistent with this analysis, public opposition to powerlines began to grow during the 1950s, with subsequent attitudinal changes paralleling economic changes. Presently, debt is a way of life for most farmers. However, unlike the 1920s and 1930s, private land is now in shorter supply and land values are much higher. Farmers are perhaps less willing to accept minimal compensation for selling easements because they need the money and want as much compensation as possible for encroachments on their land. Some of the most vociferous opposition has come from small farmers who see utility easements as symbolic of their more general economic plight and loss of control in the face of large institutions (such as big absentee-owner farms). One dramatic symptom of this historical turnaround is the fact that some of the same individuals who organized rural electric cooperatives and electrification in the 1930s were among the most vehement opponents of high-voltage transmission line construction in Minnesota in the 1970s (Casper and Wellstone, 1981).

A similar relation between economic conditions and farmer attitudes toward powerlines seems to have occurred in Canada. During the late 1940s and early 1950s, there was a shortage of power in southern Ontario due to an expanding economy (and low rainfall which restricted hydro-electricity). Power was even shut off for short periods on a daily basis. These conditions increased the pressure for more electricity, and may have created positive public attitudes toward powerlines. Now, attitudes appear to be more negative, perhaps because the economy is less growth-oriented and power is readily available (Mitchell *et al.*, 1976).

Residential landowners went through an historical pattern similar to that of farmers. As long as suburban or rural residents lacked electricity, they were probably eager to obtain the convenience of electric service, and viewed the arrival of transmission lines as a distinct benefit. However, once electricity was widespread and a new generation grew up with it as a given, the positive symbolism of transmission lines seems to have declined. Concurrently, the value attached to an unpolluted environment has increased. Opponents of powerlines have expressed this change themselves:

> There has been a very great shift in people's attitudes toward progress. People are a lot more concerned now about their air and their water and farmers about their land . . . And to build a big powerline in the name of progress I think is old-fashioned (Casper and Wellstone, 1981).

The greater willingness of younger people than of older people to pay extra for pollution-free energy (Survey Research Laboratory, 1977) undoubtedly reflects this historical change (rather than simply an age difference) in attitudes.

Description and Discussion of Key Elements

Property value effects

Elsewhere we have reviewed evidence relating to the effect of transmission lines on property values (Furby *et al.*, in press). Here, we describe people's *perceptions* of that effect, since those perceptions are likely to shape their attitudes toward transmission lines.

In a frequently cited study, Kinnard (1967) reported a survey conducted in Connecticut in the mid-1960s. A questionnaire was mailed to 838 homeowners who had bought property in subdivisions which were either intersected or abutted by a high-voltage powerline right-of-way. Of these, 46% responded with a completed questionnaire. Of the 80% of respondents who indicated that they were aware that there was an electric powerline right-of-way in the vicinity when they made their purchase, 76% said that its presence did not affect their decision to buy or the price they paid. From this, Kinnard concluded that 'only a minority' of purchase decisions are affected by a powerline right-of-way. However, the low rate of questionnaires returned renders any such conclusions quite tentative. Although a 46% response rate is good as surveys go, all that can be said from this study is that 28% (i.e., 76% of 80% of 46%) of all of the homeowners within these subdivisions affected by a right-of-way said that it did not affect their purchase decision. That leaves 72% who either said it did affect their decision, who were unaware of it, or who did not answer the questionnaire. Moreover, the sample does not include people who decided not to buy in the area because of the powerline.

In the same study, Kinnard also surveyed assessors, appraisers, builders, lenders, and realtors in the area, but again the response rate was too low for all of the groups (ranging from 25–54%) to permit any general conclusions. Of those who responded, however, a substantial majority felt that transmission line rights-of-way reduced the value of property that they intersected. For example, 71–79% of appraisers felt there would be a negative impact (the exact value depends on whether the property is a vacant or improved lot). For realtors the figures are 87–90%; for lenders, 63–76%; for builders, 62–70%. For properties abutted (rather than intersected) by a line, a smaller number, but still a majority in each group, thought that property values are negatively affected. The intersected vs. abutted distinction was smallest with realtors. Kinnard noted that both realtors and appraisers 'were particularly negative in their reactions toward the saleability or marketability of both acreage and developed residential properties either intersected or abutted by powerline rights of way' (p. 279).

Much anecdotal testimony by appraisers in the literature contradicts Kinnard's survey results (Furby *et al.*, in press). It is hard to know whether this discrepancy reflects non-response bias in Kinnard's study (with appraisers who believe that transmission lines do not affect property values being less likely to complete the questionnaire) or, alternatively, a self-selection bias in appraisers who choose to speak publicly or write about the issue.

However accurate Kinnard's conclusions are, they remind us that the perceptions of professional appraisers are of interest not only in their own right but also for their effect on other people's perceptions. Their opinions are likely to 'translate themselves into advice to potential buyers or sellers, and/or judgments about the market value of such properties' (p. 279). It is, therefore, unfortunate that there is so little systematic empirical evidence regarding appraisers' and realtors' perceptions of transmission line effects on property values, and that the existing studies are so dated.

About a decade after Kinnard's study, Mitchell *et al.* (1976) compared residents along a 230 kV line corridor in Ontario with a control group of property owners living along a parallel corridor one mile away. Both groups were asked about the line's effect on the value of the corridor residents' property. In each group, 25 people were sampled, of whom only 16 (64%) were actually interviewed for the survey (the others were not home, refused, etc.). Five of the 16 residents (31%) felt that their property's value had decreased, whereas 15 of 16 in the control group (94%) felt that the property of those living near the line had decreased in value. The very small sample size and the low response rate render these results difficult to interpret. If the group difference is real, it could mean that the non-residents are ill-informed and/or that residents are denying the effect that the non-residents perceive.

A second survey conducted two years later by the same group (Boyer *et al.*, 1978) compared residents near both a 500 kV and a 230 kV line with control groups living one mile from the transmission line corridors. Of the 25% of residents who agreed to fill out the questionnaire, approximately 70% actually returned it, resulting in 108 completed questionnaires spread relatively evenly across the four groups. In the two resident groups combined, 44% of those who had bought their property after the transmission line was constructed reported that its presence was a consideration or a strong consideration in their decision; 69% said they felt the line affected the value of their property. In the control group, 78% said that the presence of a transmission line would be a consideration in their decision to buy land. Only 43% of the control group believed that the existing line affected the value of the property adjacent to it (but 45% said they didn't know, meaning that 79% of those with an opinion felt that it affected property values). As before, non-response bias blurs the meaning of these results. Even the most liberal assumptions leave a possible range of 31–62% for whom the transmission line was a consideration when purchasing property (leaving unspecified whether the line was considered to be positive or negative).

Given the relatively limited number of studies directly addressing property value perceptions, and their rather severe methodological shortcomings, the empirical basis for understanding this issue is still very weak.

Aesthetics

The perceived physical change to the landscape is generally referred to as the 'visual impact' of transmission lines. That impact is assumed to be negative, diminishing the attractiveness of the visual landscape. Several types of empirical studies have examined the visual impact of transmission lines.

The two University of Waterloo studies described above (Mitchell *et al.*, 1976; Boyer *et al.*, 1978) included a general question about aesthetics, 'Does the appearance of the transmission line bother you?' The 1976 study found two 'yes' responses (14%) among 16 respondents. In the 1978 study, with 54 respondents, 36% objected to the appearance. About the same percentage (37%) of the control group in that study also reported that the appearance of the line bothered them (even though they did not live right next to it). Forty-eight percent of residents living next to it and 58% of those in the control group reported the line to be a prominent or very prominent landscape feature. As already mentioned, these results are subject to non-response problems.

By far the most extensive survey of people's attitudes toward the visual impact of transmission lines was conducted in 1972 by Response Analysis Corporation (see also Pohlman, 1973). Their nationwide interview study included 1962 individuals chosen to be representative of the U.S. population of adults 18 years and older. Unfortunately, their report does not indicate what response rate they achieved, making non-response bias difficult to assess. However, their sample was quite similar to the U.S. population of adults in terms of age, gender, education, and region of residence. When asked to indicate the two or three most unattractive things in their neighborhoods from a list of 14 items, 12% of the respondents cited high-voltage transmission structures, which ranked eighth behind such items as litter and trash (mentioned by 70%), poorly paved streets (49%), junkyards (48%), and telephone and electric utility poles (13%). Given that only 30% of the respondents indicated having high-voltage transmission structures in their immediate area (the percentage actually living near lines was unknown), 40% of those aware of high-voltage lines nearby considered them to be among the two or three most unattractive things about their area. This study also reported that (a) the identification of transmission structures as unattractive was most common among urban residents, followed by suburban residents, and then rural residents; (b) people with high-voltage structures in their area were more likely to consider them generally unattractive than those who did not have them nearby, and (c) steel poles were preferred to other designs by a clear majority (67%).

A number of recent studies of the visual impact of transmission lines can be found in the literature on visual analysis and resource management (Priestley, 1983a). The most common research method involves showing respondents slides of transmission lines placed in a variety of landscapes and asking them to evaluate the visual compatibility of the lines, i.e., the degree to which the lines fit in or blend with the landscape, independent of the landscape zone's visual quality. Jones and Jones (1976) used this method with citizen groups in Idaho and Montana having different orientations toward the environment including a grange, wilderness association, Rotary Club, Lions Club, historical society, and mining association. The samples were relatively small (6–38 per group) and cannot be considered representative in any statistical sense. Nevertheless, the results indicated that there is considerable agreement across groups regarding relative compatibility of various scenes that include transmission lines, suggesting that individuals use the same visual features to judge compatibility regardless of their attitudes about land use and development. The groups differed in their mean ratings of compatibility, with the 'environmental/conservation' groups giving the lowest ratings, the 'intensive land-use, groups giving the highest ratings, and the 'general business' groups falling in between. Jones and Jones interpret these differences as reflecting *attitudinal* differences in these groups' degree of acceptance of transmission facilities.

Using a slightly different approach, Brush and Palmer (1979) had a group of 30 landscape architects and planners evaluate 124 photographs representing various combinations of landforms and land use types resulting from urbanization in the Connecticut River Valley. Each respondent performed a Q-sort, sorting the photographs into seven piles from lowest to highest scenic quality. Scores were assigned from 1 to 7, and then each photograph's score was obtained by averaging across all respondents. Multiple regression analyses showed that the number of utility poles had the greatest negative effect on the rated scenic quality of seven regional landscape classes (farms, forested hills, meadows, open water, wetlands and streams, towns, and industry). There were only two other variables with significant negative effects: length of utility wires and amount of barren ground in the far distance.

A similar study by Jackson, Hudman and England (1978) asked 1500 individuals to indicate their relative preference for 72 different scenes, a number of which included transmission lines of varying types and locations. Factor analytic results suggested that the salience of powerlines accounted for about 15% of the preference rankings, making it the second most important determinant of scenic evaluations (the first factor consisted of the degree to which human structures dominated the scene).

Another potential but apparently neglected source of information on attitudes toward the aesthetics of transmission lines is how much it is worth to people to put the lines underground. We have been unable to locate any systematic empirical studies on this topic. A related source is the judgments of aesthetics experts, including artists, who have sometimes been called upon to testify in legal proceedings designed to determine whether the costs of undergrounding are warranted by the gains in scenic beauty (e.g., Gussow, 1977). For the past several decades, decisions regarding aesthetics in design and siting have relied principally on landscape architects' professional judgments (e.g., Crowe, 1958; Johnson, Johnson and Roy, 1970). However, the assumptions underlying these professional judgments have never been empirically validated.

It is clear from this review that most of these studies have found negative aesthetic effects of transmission lines in various settings, but they have rarely examined the role of specific structural and design characteristics of a line in determining people's reactions to it.

Environmental concerns

In discussing the determinants of attitudes toward transmission easements, we emphasized the importance of an historical perspective in order to fully understand current public concerns and opinions. The role of environmental quality issues must likewise be examined in historical context. By 1970, the public's attitude toward the environment had clearly changed. 'They resent the land being unnecessarily disturbed, whereas a few years ago no one cared ... They now sometimes get violent when trees are cut down' (D. L. Sweet [the director of AT&T's right-of-way activities in five midwestern states], 1970, p. 39). This shift in attitudes toward the environment reflects a more general change, with the importance of efficiency and productivity decreasing and that of aesthetics, pollution, health, and safety increasing (Katz, 1971). The National Environmental Policy Act institutionalized this change by requiring that non-economic impacts be considered in decisions affecting the environment.

The following two subsections focus on studies of the role of environmental concerns in determining people's attitudes toward transmission lines. Our objective is not to review the literature on environmental effects or health effects *per se* (see BPA, 1982; Morgan *et al.*, 1983), but rather to examine people's perception of those effects and their resulting attitudes toward transmission lines.

Noise

High-voltage transmission lines produce a noise due to corona, which is a random high-energy discharge from the conductor. Its nature and magnitude vary depending upon weather conditions and type of current (AC or DC). Its aversiveness to the human ear is just beginning to be studied.

In a very small survey, Busby *et al.* (1974) interviewed 18 farmers along 765 kV lines in Ohio. Although never asked specifically about noise, five of the 18 mentioned it in the course of the interview. In a larger study, Fidell *et al.* (1978) asked 270 people living adjacent to 230 kV and 500 kV transmission lines to evaluate the noise created. Interpretation of their data is complicated by their failure to report the response rate. Of those who agreed to be interviewed, 42% reported hearing noise from the transmission lines. However, only 2% spontaneously mentioned powerlines or transformers as a source of noise annoyance. Measurements of noise levels at the interview site revealed about ten percent more people being annoyed than would be expected from previous studies on the general relationship between noise level and annoyance. Other studies (e.g., MacLaren, 1981) have noted the unique qualities of corona noise and the need for further research on how people experience it in the vicinity of transmission lines.

Biological effects

Another major environmental concern is the effect of transmission lines on plants and animals. Concern over interference with birds' migratory paths was recently a significant source of opposition to a 500kV line in southern Oregon (Envirosphere, 1983). Potential electrical field effects on other animals (such as dairy cows) have been of concern as well, and are likely to take on increasing significance as human health effects become a more prominent issue (see below).

A 1980 survey in Minnesota (Genereux and Genereux, 1980) studied landowners' perceptions of the animal and human health effects associated with a ±400 dc kV line that crossed their properties and had created enormous controversy during the 1970s (Casper and Wellstone, 1981). After a mailed questionnaire produced a return rate of only 19% (from a total of 115), it was supplemented by phone administration to another 233 persons and by results from a similar questionnaire administered by a state representative. The transmission line protest group (GASP) produced another 56 completed questionnaires. Combining these data raises obvious methodological problems. Furthermore, sampling procedures were questionable, as evidenced, for example, by the fact that 37% of the respondents did not even live on the land they owned that was crossed by the powerline. Twenty-nine percent of the respondents said their livestock appeared to suffer adverse effects which they attributed to the powerline (another 26% said they didn't know whether their livestock had suffered such effects). The most frequently reported symptoms were breeding problems, aborted or deformed offspring, stress or nervousness, and changes in milk production. Twenty-two percent of respondents said they had noticed unusual behavior in their animals (nervousness, restlessness, ill temper). Thirty-one percent said they believed that the powerline had affected nearby wildlife.

While there remains great uncertainty as to the effects of electric and magnetic fields on animals, it is a topic that is receiving increasing attention in the media (e.g., Squires, 1985) and as such we might expect greater public concern about it in the future.

Human health and safety issues
In the 1978 University of Waterloo survey (Boyer *et al.*, 1978), 35% of those living along a line were aware of controversy over possible health effects, while 32% of the control group (living a mile away) said they were aware. Fifty-nine percent of the on-line group said they were concerned or very concerned about health effects (apparently some people are concerned even though they aren't aware of any controversy). Genereux and Genereux (1980) found that 35% of all landowners whose land was crossed by a DC transmission line (and 48% of those actually living on that land) reported suffering adverse health effects that they attributed to the powerline. The most frequently reported symptoms were headaches, respiratory problems, fatigue, and stress. As mentioned above, serious sampling and data collection difficulties make it unwise to draw any general conclusions from either of these studies.

An interesting secondary analysis of the Genereux and Genereux study (Gatchel, Baum and Baum, 1981) discusses whether the symptoms reported can be attributed to stress (resulting from opposition to the line, worry over the health effects, etc.) rather than to electrical effects *per se*. Evidence for analogous stress effects has been found with residents in the Three Mile Island area (Baum *et al.*, 1981). Gatchel *et al.*, concluded that the type and frequency of symptoms might indeed be due to stress. This possibility makes the study of *perceived* health effects especially important. Not only do such perceptions shape attitudes toward transmission lines, but they may actually produce their own health effects.

Perceptions may also shape jury decisions. A dramatic demonstration of the importance of perceived health risks is the 1985 ruling by a Texas jury ordering Houston Lighting and Power Company to pay more than $25 million in punitive damages to a local school district for 'reckless disregard' of children's health in siting a 345 kV transmission line on school property ('Cancer Risk,' 1985). The jury also ruled that if the utility did not remove the powerline, it would have to pay the costs of relocating the school buildings on the property (estimated at $64 million).

In contrast to Genereux and Genereux, who asked people what symptoms they attributed to the transmission line, Nolfi and Haupt (1982) surveyed health effects along a ±400 DC kV line in California without mentioning the line. They interviewed 128 households, representing a response rate of 62%. These households included 438 individuals, 245 of whom were within 0.14 miles of (but not always directly adjacent to) the transmission line ('powerline group') and 193 of whom were 0.65–0.85 miles from it ('control group'). People were asked how many days during the previous two weeks they had experienced nine specific symptoms. None of the symptoms studied was reported more frequently by the powerline group than by the controls, leading the authors to deny any relationship between proximity and health problems. However, they did find a statistically significant relation between two symptoms (sore or dry throat, and stuffy nose or respiratory congestion) and whether or not one lives directly adjacent to the line (but they dismiss this as being questionable because of the small frequencies involved). The authors themselves point out that their sample was too small to draw conclusions about infrequent health effects, and that some potential symptoms (e.g., miscarriages, cancer) were just not mentioned. Furthermore, the transmission line was not operating for most of the two-week period to which the symptom frequency questions referred (Banks, 1984).

As long as the health effects of powerlines remain ambiguous, people will undoubtedly be tempted to see powerlines as a source of health problems.

Psychometric analyses

One broad strategy for studying the perception of health and safety is to develop a taxonomy for hazards that can be used to understand and predict responses to their risks. A taxonomic scheme might explain, for example, people's extreme aversion to some hazards, their indifference to others, and the discrepancies between these reactions and experts' opinions. The most common approach to this goal has employed the *psychometric paradigm* (Fischhoff *et al.*, 1978; Slovic *et al.*, 1984, 1985) which uses psychophysical scaling and multivariate analysis techniques to produce quantitative representations of hazards and people's perceptions of them. In the psychometric approach, people make quantitative judgments about the current and desired riskiness of diverse hazards and the desired level of regulation of each. These judgments are then related to other judgments about the hazard's status on various characteristics that have been hypothesized to account for risk perceptions and attitudes (e.g., voluntariness, dread, knowledge, controllability, catastrophic potential, threat to future generations, etc.).

Many of the risk characteristics are highly correlated with each other, across a wide domain of hazards. For example, voluntary hazards tend also to be controllable and well known. Analyses of these interrelationships suggest that the broader domain of characteristics can be condensed to two or three higher-order characteristics or factors. Figure 2 illustrates the factors that emerged from one study of 81

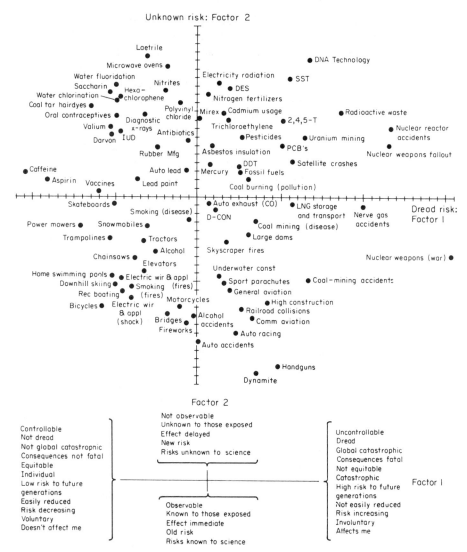

FIGURE 2. Hazard locations on Factors 1 and 2 derived from the interrelationships among 18 risk characteristics. Each factor is made up of a combination of characteristics, as indicated by the lower diagram. (Source: Slovic *et al.*, 1985).

hazards, each rated by college students on 18 risk characteristics. Two main factors were obtained. Hazards located at the high end of the vertical factor (e.g., DNA technology, various chemicals) tended to produce risks that were judged new, unknown, and delayed in effect. Hazards at the other extreme of this factor (e.g., dynamite, automobile accidents) had the opposite characteristics. High scoring hazards on the horizontal factor (e.g., nuclear reactor accidents, nuclear weapons effects) were associated with risks that were judged to be uncontrollable, dreaded, catastrophic, fatal, and inequitable. Hazards low on this factor (e.g., caffeine, power mowers, skateboards) were seen as causing injuries rather than fatalities, to single individuals. We have labelled the vertical factor as *Unknown Risk* and the horizontal factor as *Dread Risk*. This factor structure has been found to be similar across groups of laypersons and experts judging large and diverse sets of hazards.

Research has shown that lay people's risk perceptions and attitudes are closely related to the position of a hazard within the factor space (Slovic *et al.*, 1982). Most important is the factor 'Dread Risk': the higher a hazard's score on this factor (i.e., the further to the right it is in the factor space), the higher its perceived risk, the more people want to see its current risks reduced, and the more they want to see strict regulation employed to achieve the desired reduction in risk. In addition, the informativeness or 'signal potential' of an accident or mishap, which appears to be a key determiner of its perceived seriousness, is systematically related to both Dread Risk and Unknown Risk factors (Slovic *et al.*, 1984). As a result, mishaps associated with hazards located in the upper-right quadrant of the space are likely to receive extensive media coverage and to produce 'ripple effects' that are very costly to the agency or industries responsible for their management.

A psychometric study by Slovic *et al.* (1985) included 'electricity radiation from high-voltage transmission lines among a set of 81 hazards judged by a group of college students. As shown in Figure 2, this hazard was judged to be high on the Unknown Risk factor and a little above average on the Dread Risk factor. The upper-right quadrant of the factor space, where this hazard fell, is generally populated by hazards for which public concerns are high and acceptance is a volatile matter. The results of this study are at best suggestive, due to the non-representative nature of the sample. Furthermore, the (inappropriate) use of the term 'radiation' in describing the hazard from high-voltage transmission lines may have enhanced people's concerns.

A more elaborate psychometric study, focusing specifically on 50/60 Hz electric and magnetic fields was conducted by Morgan *et al.* (1985). The participants in this study were 116 alumni of Carnegie-Mellon University, many of them engineers. They were asked to evaluate the potential hazards from 50/60 Hz fields produced by high-voltage transmission lines and by electric blankets. They also evaluated the hazards accociated with 14 other substances and technologies. The sixteen hazards were rated on nine risk characteristics used in studies by Slovic *et al.* (1985). Additional questions were asked about the significance of the risk posed by each potential hazard and the adequacy of existing measures for controlling such risks.

The results showed that these individuals viewed exposure to electric and magnetic fields from large transmission lines and from electric blankets as among the least risky of the 16 known and potential hazards they considered. Forty-eight percent judged electric blankets as least risky of the hazards and 19% judged transmission lines as least risky.

On the basis of the ratings for transmission lines on the nine risk characteristics, the similarity between this potential hazard and hazards investigated in other studies could be assessed. Hazards that appear to be most similar in character to transmission lines in two such studies (labelled A and B) are shown in Table 1. The similar hazards from Study A (e.g., space exploration and non-nuclear electric power) are not ones whose risks are particularly controversial. However the similar hazards from Study B are mostly chemicals, which are a source of much concern these days.

Factor analysis of the risk ratings produced a space (Figure 3) similar to that of previous studies (see, e.g., Figure 2). Electric and magnetic fields from large power-lines again appeared in the high unknown, high dread portion of the space. The risks from electric blankets were judged relatively unknown but much less dreaded.

TABLE 1

Ten hazards in studies conducted by Slovic et al. (1985) most similar to transmission line fields

Study A		Study B	
Hazard	*'Distance'*	*Hazard*	*'Distance'*
Earth orbiting satellite	1·75	Nitrogen fertilizers	1·35
Space exploration	2·24	Polyvinyl chloride	1·68
Solar electric power	2·94	High tension lines	1·87
Non-nuclear electric power	3·07	Cadmium usage	2·01
Fossil electric power	3·12	Airborne lead from autos	2·21
Food coloring	3·17	Chlorination of	
Hydroelectric power	3·18	drinking water	2·30
Food irradiation	3·22	Trichlorethylene	2·45
Food preservatives	3·25	Nitrites	2·46
Water fluoridation	3·29	Mercury	2·47
		Mirex (insecticide)	2·49

Source: Morgan *et al.* (1985).

Note: The distance measure was $\sum_{i=i}^{9} d_i^2$ where i is the index running over the nine risk

characteristics and d_i is the difference between ratings for the ith characteristic. The smaller the difference, the greater the similarity. There were 90 hazards in Study A; 81 hazards in Study B.

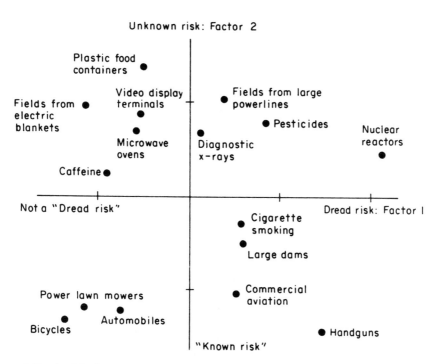

FIGURE 3. Result of the factor solution for the sixteen known or potential hazards addressed in the study by Morgan *et al.*, 1985.

The location of large powerlines in the factor space allows an informed conjecture about how well-educated people are likely to respond to these potential hazards. It suggests that such people are likely to want significant, but not severe, regulatory control of transmission lines. Should 'events' occur that imply health problems due to transmission lines, they will likely be seen as having high 'signal value', which would produce considerable media publicity and a great deal of concern. It would obviously be useful to have similar information for segments of the population other than graduates of Carnegie-Mellon University.

Economic benefits

The elements we have considered thus far generally contribute to a *negative* attitude toward transmission lines, but it must not be forgotten that various benefits obtained from electricity tend to have the opposite effect. As discussed in the Historical Perspective section, the comfort and convenience that electrification brings to our lives, and the economic efficiency it provides to business and industry, can be powerful determinants of attitudes toward powerlines. These benefits can easily be overlooked in a society that has become so accustomed to uninterrupted and un-limited electricity everywhere that many people fail to realize that alternatives to transmission lines included brownouts, blackouts, and other disturbing scenarios. Changes in the salience of electricity benefits (e.g., historically, or after a major blackout) might be expected to have quite significant effects on attitudes toward transmission lines. However, there have been no empirical studies as yet that address this issue.

Equity issues

Transmission lines are just one example of a number of development projects that are frequently objectionable to those living near them. Popper (1981) has dubbed such projects Locally Unwanted Land Uses or LULUs (which include airports, highways, nuclear power plants, strip mines, prisons and military installations). LULUs usually offer benefits to some locality or region other than the one suffering the costs. It is this inequitable apportionment of costs and benefits that often triggers public dissatisfaction and conflict, especially among those who feel they are suffering a disproportionate share of the costs.

Equity and fairness are increasingly common issues in discussions of many public policy decisions including those dealing with power generation and transmission. Individual landowners and even whole communities sometimes feel unfairly imposed upon, forced to involuntarily shoulder the burden of providing benefits to others (usually urban populations). A Minnesota farmer put it this way: 'Why do you have to avoid towns? . . . Why not put it [the transmission line] right over all the towns. They're the ones that are getting the benefits' (Casper and Wellstone, 1981, p. 71).

There is, of course, no perfect solution to the inequity problems. It may be necessary sometimes for a few individuals to sacrifice more than their fair share for the common good. Furthermore, judgments about fairness have multiple determi-nants which are only partially understood (Furby, 1986). In any case, the perception of unfairness can be a source of divisiveness and social disruption. Methods of minimizing the degree to which people feel unfairly imposed upon are therefore of interest. One such method is to use decision-making procedures in the siting process that are perceived to be fair (see later section on Process Characteristics), as the outcomes of those procedures are then more likely to seem fair as well (Tyler, 1984).

Another method for reducing the ill-will engendered by the perceived inequities that sometimes result when siting and constructing transmission lines might be simply to increase monetary compensation offered to the landowners. It may well be that the prevalent practice of limiting compensation amounts to the 'market value' of the affected property (see Furby, *et al.*, in press) leaves most owners feeling unfairly used by the rest of society, as they usually are not wanting to sell their property when it is taken. A larger monetary compensation might reduce the feeling that they have suffered a net loss and thus made an unfair sacrifice, while at the same time costing the beneficiaries (i.e., ratepayers) relatively little. In addition, monetary compensation could be considered for those who do not own land used by the line but who nevertheless are affected by its presence.

We have found no empirical tests of the degree to which compensation rates affect perceived inequities. However, at least one utility (Ontario Hydro) has assumed the effect to be important, as evidenced in their doubling the compensation rate (from $5 to $10 per pole in a field) in the face of dissatisfied landowners in the early 1950s. An excellent recent discussion of the strengths and weaknesses of compensation in the context of siting locally unwanted facilities is provided by O'Hare *et al.* (1983). These investigators argue strongly for the value of compensation in promoting community acceptance of such facilities. Their discussion of five case studies of siting problems provides a valuable start towards understanding the conditions necessary for compensation to be effective in achieving public acceptance of a facility.

Symbolic meaning
A salient determinant of opposition to lines seems to be their symbolic interpretation as an invasion of one's home territory. In his analysis of the Minnesota conflict, Priestley (1983*b*) points out that

> What the protesters seemed to be most concerned about was not the powerline's tangible effects, but its symbolic effects as an intrusion into their turf. The most obvious themes in the public statements of those active in the powerline protest concerned their strong sense of identification with the land, and their sense that the powerlines would constitute both a real and a symbolic violation (p. 14).

As one Minnesota farmer put it, 'If you'd've told me I'd be fighting the state like this, I'd've said, "You're nuts!" But we've been invaded!' (Wasserman, 1979, p. 39).

The attachment to one's personal property can be the product of a number of factors. Interviews with farmers protesting the Minnesota line suggested that ancestral ties, commitment to working on the land, and the hard work associated with the land all played a role (Casper and Wellston, 1981; Priestley, 1983*b*). In addition, property itself is often both a symbol and a means of exerting control and power in one's life (Furby, 1978). The Minnesota farmers had already felt the squeeze of the large corporate farms. The intrusion of the transmission line exacerbated their increasing sense of a loss of power and control. As one farmer put it, 'The powerline represents control, something else in control of our lives . . . We have been feeling this control over our lives coming' (Casper and Wellstone, 1981, p. 303–304).

The presence of transmission lines can also constitute a continual reminder of all the other issues surrounding their construction. This aspect was also articulated by the

Minnesota farmers: '. . . every time I see the towers, every time I walk in the fields, it kind of brings back all the memories of fighting the thing, and in a way every time I see it, I feel more bitter' (Casper and Wellstone, 1981, p. 120). In this respect, the physical presence of the line becomes a source of continual stress because of its symbolic meaning. As the Gatchel *et al.* (1981) study concluded, some of the health effects reported by people living near the completed line may well be due to such stress.

Information and knowledge

Naturally, what people know about a particular issue affects their attitudes (McGuire, 1969). Two aspects of such knowledge are of particular interest for attitudes toward transmission lines. One is exposure to current-event information about transmission lines, such as regularly appears in the press (e.g., utilities' plans for new installations, reports of public opposition to controversial lines). The second is exposure to scientific information about such issues as the economics of alternative types of installations, or the biological and health effects of powerlines. Empirical studies specifically linking the amount of exposure to such knowledge and attitudes toward transmission lines are, however, few in number.

The public's knowledge of electric power planning and operations was examined to some extent in the Response Analysis Corporation's (1972) nationwide survey. It reported that 63% of the adults surveyed felt that either none or only a few additional lines would be needed in the next ten years (another 23% said that they didn't know how many lines would be needed). This was at a time when the industry experts were forecasting large increases in transmission line mileage. Half of those surveyed didn't know if there was state or local control over the type of high-voltage transmission structures that are put up in their community. These results clearly suggest a lack of general information about transmission lines. Studies of how individual landowners deal with the utilities during the planning and construction of a line also have found that people are generally very poorly informed about the regulatory process with regard to transmission lines (Young, 1973; Casper and Wellstone, 1981; Gale, 1982).

A study by Tichenor *et al.* (1980) reported that high levels of knowledge about a controversial transmission line in Minnesota did not necessarily accompany a particular attitude, suggesting that raising knowledge level per se might not be expected to have a predictable effect on attitudes. However, at least one case of a utility's deliberately attempting to increase public knowledge about a proposed transmission line suggests that such an effort may reduce negative reaction and opposition (Rydant, 1984).

The provision of information has been examined more specifically with respect to health effects in the aforementioned study by Morgan *et al.* (1985). Besides characterizing people's perceptions of risks from electric and magnetic fields, this study also examined experimentally the effects of information on attitudes and perceptions. After producing the data summarized in Table 1 and Figure 3 above, participants in the study received detailed but non-technical information regarding:

1. The nature of electric and magnetic fields;
2. What is known about possible undesirable health impacts of exposure to 60 Hz fields;
3. How the fields from transmission lines compare in strength with other Hz fields (this information was mostly pictorial).

After receipt of this information, respondents again evaluated the risks from electric and magnetic fields produced by transmission lines and electric blankets. Provision of information moved the evaluation of transmission lines further into the upper-right quadrant of the factor space (Figure 3). Possible effects of electric and magnetic fields from transmission lines were seen as more dreaded, less equitable and less well known to science. Similar changes occurred in the characterization of the risks from electric blankets. For both transmission lines and electric blankets, provision of information also increased people's concerns that existing control measures were not adequate and increased the tendency to 'feel sure that this is a risk.'

Though few in number, these studies suggest that people may not be very well informed about powerlines, and that providing factual information can alter their attitudes.

Process characteristics
The nationwide survey conducted by Response Analysis Corporation (1972) described above included several questions pertaining to people's perceptions of the decision-making procedures used by electric companies. In response to one question, 68% of the respondents said the companies were either somewhat or greatly concerned about the appearance of transmission structures. Forty-six percent answered 'yes' to: 'If you and your neighbors opposed new transmission structures in your area, do you think the electric company would consider your feelings, or not?' (20% said they were unsure). Unfortunately, drawing conclusions from these results is somewhat problematic, given that the non-response rate for this study was not reported.

The public appears to be quite concerned about being left out of the transmission line planning process. In years past, this perception was probably accurate. Utilities sometimes tried to conceal their plans for a line before construction began, in order to minimize both the possibility of political opposition and the cost of land acquisition (Porter, 1980). More recently, more utilities have realized the value of involving the public early in the planning process, both because such involvement is sometimes mandated (e.g., *Guidelines for Linear Development*, 1977), and because there is evidence that public opposition is less likely when people are informed and involved from the beginning (DeVore, 1969; Ewald, 1969; Fraggalosch, 1980). Such involvement is also in keeping with the shift in public values with respect to the environment and the political process. 'In a society that aspires to be free and open, such judgments can be made only through discussion, argument, persuasion, contention, adjustment and interaction among the individuals and groups with a stake in the outcome' (Katz, 1971, p. 196). The utility industry has often been ill-equipped to deal with non-engineering problems such as this shift in values and the resulting public opposition to transmission lines. Kipp (1969) pointed out that highway agencies faced a similar situation in the mid-1950s when the federal government authorized the construction of thousands of miles of limited-access highway to meet increasing transportation demands. Public resistance unexpectedly appeared once construction began, and only a collaborative effort between social scientists and transportation engineers in the form of detailed community planning studies managed to reduce the community–highway conflict.

If opposition is prompted by feelings of being ignored, an obvious antidote is including the public in the process. Means for doing so include environmental impact hearings, citizens' advisory councils (e.g., Pennsylvania Power & Light, 1982), and community approval requirements. These, of course, require considerable expenditures of time and money on the part of utilities (and often by private individuals and public entities as well). There may, however, be much to gain from such expenditures. Studies of procedural justice (Tyler, 1984) suggest that increased public participation generally enhances perceptions of process fairness. Furthermore, when the public is forced to consider the various alternatives (e.g., more small generating facilities and consequent increased fuel transportation costs; expensive underground lines; inconvenient and costly energy conservation measures), it may be more willing to live with the shortcomings of whichever option is selected, after the shortcomings of alternatives have also been made salient.

An alternative to including the public in the decision-making process is including a representation of its values. Relatively detailed techniques have been developed to identify those values with respect to transmission line routing. One example of this approach is a study by Hendrickson *et al.* (1974) which applied a method based on multi-attribute theory to the problem of siting transmission lines. Initially, the investigators identified criteria believed to influence people's evaluations of routes: aesthetic, economic, health and safety, water quality, air quality, cultural and recreational, and animal and health impact. Representatives of the public then assigned weights to these criteria, either explicitly or implicitly by rating various viewscapes showing different types of transmission line routing. Technical experts judged the impact of alternate routes for each of the criteria. All of this resulted in a rating score for each route based on both technical and public judgment. Whether such an approach can satisfy the public remains to be seen. In a study evaluating the use of a multi-attribute methodology for assessing visual impacts only, public reaction was fairly critical of this approach, stressing that value judgments must be made on an individual case basis, and that the public must be allowed to question any and all assumptions or preconceptions (Enk *et al.*, 1981).

Public participation is becoming an issue not only for transmission line siting but also for need forecasting and project justification. Case studies of public opposition to specific transmission lines (e.g., Fraggalosch, 1980) show increasing public disagreement with the utilities' forecasted need for more power, and increasing public demand for participation in longer-range planning and energy decisions. In addition to the desire to be counted, the distrust seems due to the errors in energy need projections made during the 1970s. There is also a significant portion of the population that favors smaller-scale technologies that eliminate the need for large transmission lines. Even utilities that have incorporated the public into their siting procedures have thus far strongly resisted any public involvement in the larger decision about the need for new energy facilities (Fraggalosch, 1980). Such participation may only be achieved through state and local regulatory mechanisms requiring public input in the assessment of energy needs (e.g., Minnesota Environmental Quality Board, 1982).

Public attitudes and perceptions are reflected not only in individual citizen's opinions, but also in representative public bodies. Sweet (1970) reported a questionnaire designed to assess how such public bodies view the procedures used by companies involved in right-of-way management including telephone, gas, water, and electric companies. The questionnaire was mailed to a number of public agencies around the country, but neither the sampling procedure nor the response rate was reported, rendering the results of questionable utility. Given this caveat, the power companies generally had a reasonably good image with public agencies. For example, 93% of the agencies responding said the power companies' initial requests for siting authorization were made in a reasonable manner; 99% said the utility worked in accordance with the agency's specification; 94% said that the utility did not disrupt normal community activities; 75% rated power companies' images as 'good.' The power companies received somewhat lower marks in several other respects: 27% of the agencies said that utilities sometimes failed to obtain agency authorization before starting work; 26% said the utility's follow-up after completion of the work was only fair or poor; 11% said they had received unfavorable reaction or comments from the community as a result of utility company work. Sweet summarized his own conclusions from the study rather emphatically:

> It seems to me the message is loud and clear—many of us [utilities] are not satisfying the various public agencies.
> – We have some problems in the area of applying for permits or licenses.
> – Our construction leaves something to be desired.
> – They want to be in on the planning of our new facilities.
> – Finally, they are asking for more and better liaison.
> I personally think we had better listen to what they are telling us and take action, otherwise tomorrow there may be legislation passed which we don't like (p. 39).

Given the methodological limits of this study and the pivotal role of public officials in creating and coping with public agencies, there is a need for more research of this type.

Attitudes toward the siting and construction process are represented in Figure 1 at the same level as attitudes toward the line themselves. Both are influenced by a number of determining elements, and both can directly affect acceptance of or opposition to a proposed line.

The dynamics of opposition and conflict

The various elements discussed thus far appear to play a significant role in determining attitudes toward transmission lines. However, while knowledge of attitude determinants is important, it does not tell the whole story when it comes to understanding when and how public opposition will be manifested. A large body of research has demonstrated that the relation between attitudes and behavior is not always a close one. There are clearly factors other than attitudes that play a role in the development of community conflict. Those factors are often difficult to define, but they generally relate to the dynamics of how conflict escalates or dissipates.

Tichenor *et al.* (1980) attempted to analyze the role of the press in communicating information about community conflict over high-voltage transmission lines. From their study of the Minnesota case, they concluded that it is the organization of opposition groups that is the primary ingredient in the development of conflict rather than the manner in which the media presents the issue. Although the news media can affect any opposition group's success by the amount of coverage that they give it, they do not create the malaise.

Some have claimed (e.g., Mitchell *et al.*, 1976) that opposition dissipates over time, and thus that residents in an area traversed by a transmission line may oppose it during the planning and construction phases but eventually will adapt to it. The study by Boyer *et al.* (1978) sheds some light on this issue by comparing the attitudes of the respondents in their Ontario survey at the time the first line was constructed to their attitudes two years later when a second line was proposed. Unfortunately, most of the on-line sample had not lived on their property prior to construction of the line, so that only 23 people provided responses to the question about their attitude toward construction of the original line. Of those 23, 42% had been opposed or very opposed, while only 13% said they had been favorable or very favorable at the time of construction. For the control group (living approximately one mile away), 11% said they had been opposed or very opposed while 32% said they had been favorable or very favorable to construction of the original line. A second line was proposed for construction two years after the first (1973) in the same 500 kV corridor. The on-line group reported a fairly constant level of opposition across this two-year time period: 45% reported that they were opposed or very opposed and only 17% said they were favorable to addition of the second line. Furthermore, when asked which alternatives they preferred for increasing electric capacity, 28% of the on-line group chose a new corridor rather than the addition of a second line to the existing corridor (while only 5% of the control group did). Thus, those living on land traversed by a transmission line are more opposed to it at the time of initial construction than those living at some distance, and several years later they remain as opposed as they were before to the construction of a new line.

One temporal pattern that has emerged in several cases of serious conflict over transmission lines is a shift from little or no mention of health effect issues during the early stages of opposition to an increasing focus on them as the conflict continues. Anecdotal evidence suggests that this is not necessarily the result of increased education about or concern over health problems *per se*. Rather, those opposing the line learn that negative health effects carry more weight in regulatory and legal battles than any of their other concerns such as aesthetics and property values.

A better understanding of the dynamics of conflict is important to minimizing future conflict over transmission lines. That understanding can probably be obtained best only from detailed case studies such as the one conducted in Minnesota.

Discussion and Conclusions

We have seen that there has been considerable research related to public attitudes toward transmission lines spanning a period of several decades. However, many of the individual studies are of limited validity or generalizability, due to small or non-representative groups of respondents. Also, we have seen that attitudes toward transmission lines have shifted significantly over time, making many of these studies seriously out-of-date. Moreover, most studies have focused on a single element of the picture (aesthetics, health effects, property values, etc.) without attention to the potential interactions among all of the elements.

As a result, the cumulative knowledge provided by all of these studies is not sufficient to allow precise statements about the organizing scheme presented in Figure 1. The literature does seem to indicate that each of the elements of the framework in Figure 1 has some importance in determining perceptions and attitudes. We do not, however, know enough to predict which of these elements will be most problematic in a particular siting situation. Issues that cause great concern in some siting controversies often are less salient in others. Indeed, each transmission line siting presents a unique combination of characteristics, and it is unlikely that we will be able to predict exactly how the affected public will react with respect to all the relevant elements.

Because of the specificity of each transmission line case, further research attempting to establish *general* principles regarding certain of the elements would seem to offer diminishing returns. Aesthetic concerns will vary with the particulars of terrain and land use, and the research to date suggests that there is no single or best aesthetic solution. Rather, the public wants to be involved in the aesthetic decisions for any given siting (Enk *et al.*, 1980). Likewise, the effects of symbolic meaning, economic benefits, and the dynamics of opposition and conflict vary considerably from case to case. While these elements are important, they may not be subject to generalizations that can be applied across multiple sitings nor used to predict the course of public attitudes to future sitings. They are probably best examined in the context of individual case studies.

There are, however, several of the attitudinal determinants we have discussed for which useful generalizations seem possible, and further research on these elements would seem fruitful. One is the effect of transmission lines on property values and public perception thereof. Property value effects are relatively easy to quantify and study empirically, and people's attitudes toward lines are influenced by how they think their pocketbooks are affected. We have outlined elsewhere a suggested research agenda in this area (Furby *et al.*, in press).

A second area is the perception of health and safety effects of transmission lines. In recent years, this issue has received increasing attention in public debate, and we suspect it will be a critical determinant of the acceptance or rejection of transmission lines in the future. A modest beginning has been made in research on this topic, but considerably more could be learned.

Most importantly, perhaps, we need to build upon research from other domains that point to the key role of process or procedural considerations in determining the nature and course of environmental conflicts (Daneke *et al.*, 1983). This is consistent with many classic studies in industrial psychology demonstrating that allowing workers to participate in problem solving and decision making produces greater acceptance and more efficient implementation of the resulting decisions (Vroom, 1969). A surprising number of those who have investigated the transmission line siting and property evaluation process state that more attention should be given to enlarging the scope and role of public participation. This was also acknowledged by representatives of the utilities and other public agencies with whom we spoke. It would appear that open discussion and increased public participation might avoid a polarization of forces, in which individual owners are pictured as selfish, greedy, and impeding social progress while utilities and public representatives are seen as high-handed and insensitive.

We recommend that future research focus on exploring the interactions among multiple elements of the conceptual framework as they relate to process issues. For example, a study that would have both theoretical and practical importance would examine the influence of process considerations on the interpretation of information about health effects. Recall that the study by Morgan *et al.* (1985) showed that information about health effects of transmission lines led to somewhat heightened risk perception and increased concern about health effects. Extrapolating from research showing that lack of control is associated with high risk perception and low acceptance of risk (Renn and Swaton, 1984; Slovic *et al.*, 1984, 1985), we would predict that the increase in concern observed by Morgan *et al.* would be even greater if this information were presented within the context of a process that was not perceived as democratic and genuinely supportive of public participation. On the other hand, we would predict that the same information, presented within the context of a more participative process, would not lead to increased concern—in fact, concern might well decrease. These predictions could be tested rather directly by replicating the Morgan *et al.* study within the context of two (hypothetical) siting processes, one incorporating public participation, the other providing little opportunity for public input.

A related aspect is that many individuals may be opposed not just to the transmission line but to the power project itself; this is particularly likely if electricity is to be produced by a nuclear plant. If attitudes toward the transmission line could be effectively separated from those toward the project (or big government, or a specific utility), the tensions surrounding negotiations might be considerably eased. Significant changes with regard to public participation, whereby individuals could have additional input into the planning process prior to the selection of alternative corridors, might help considerably in this regard.

More generally, we need research aimed at describing how to create an acceptable siting process. Citizen participation in the planning and siting process can take many forms, and careful consideration should be given to the various effects of each. Innovations in citizen participation over the past decade are beginning to provide an evidentiary base from which we can draw in assessing the relative merits of different participatory arrangements, but the literature to date is sketchy and anecdotal (Ducsik, 1981), and a more organized empirical effort would clearly be desirable (e.g., Delli-Priscoli, 1983). At present, only rough guidelines are apparent. The process should involve all affected stakeholder groups genuinely and early enough so they can exert some influence over the alternatives that are being considered. Edwards and von Winterfeldt (1987) have developed a method for identifying relevant stakeholders and helping them articulate their values and concerns in a way that can be useful for policy makers. Implications of their technique for transmission line siting would be worth exploring. A good information and education program is essential to the process, but such a program must be a two-way affair, in which each party attends to and respects the views of the other parties.

References

Ackerman, B. A. (1977). *Private Property and the Constitution.* New Haven: Yale University Press.

Anon. (1985, Nov/Dec). Cancer risk cited in $25 million power line award in Texas. *Microwave News*, pp. 5, 8, 9–11.

Anon. (1983). Envirosphere Co. *Eugene-Medford 500 kV Transmission Line: Environmental Impact Statement.* Washington, DC: U.S. Department of the Interior.

Anon. (1977, March). *Guidelines for Linear Development.* British Columbia: Environment and Land Use Development.

Anon. (1970). Johnson, Johnson, & Roy, Inc. *Transmission and Distribution Rights of Way Selection and Development.* Jackson, MI: Consumer's Power Company.

Anon. (1981, December). MacLaren Plansearch Inc. *A Report of the State of the Art of Environmental Effects of Transmission Lines.* Ottawa, Ontario: Canadian Electrical Association.

Anon. (1982, December). *Findings of Fact, Conclusions and Order: In the Matter of Considera-tion of the Adequacy of the Construction Permit for the CPA/UPA 400 kV Direct Current High Voltage Transmission Line to Protect Public Health and Safety*. St Paul, MN: Min-nesota Environmental Quality Board.

Anon. (1984, January). The national grid: Where is it now? *Ruralite*, 3–4.

Anon. (1982). *Land Use Policies and Principles*. Allentown, PA: Pennsylvania Power and Light.

Anon. (1972). *Steel Poles Versus Lattice Structures: Public Attitudes Toward High-Voltage Transmission Structures*. Princeton, NJ: Response Analysis Corporation.

Anon. (1977, October). Survey Research Laboratory, University of Illinois. *Public Reactions to Wind Energy Devices*. Washington, DC: National Science Foundation.

Anon. (1981, February 3). Yankton balks at MANDAN. *Argus Leader*.

Banks, R. S. (1984). On effects of high-voltage transmission lines [Letter to the editor]. *American Journal of Public Health*, **74**(9), 1042.

Baum, A., Gatchel. R. J., Fleming, R. and Lake, C. R. (1981). *Chronic and Acute Stress Associated with the Three Mile Island Accident and Decontamination: Preliminary Findings of a Longitudinal Study*. Washington, DC: U.S. Nuclear Regulatory Commission.

Bonneville Power Administration. (1982). *Electrical and Biological Effects of Transmission Lines: A Review*. Portland, Oregon.

Boyer, J. C., Mitchell, B. and Fenton, S. (1978). *The Socio-Economic Impacts of Electric Transmission Corridors: A Comparative Analysis*. Waterloo, Ontario: University of Waterloo, Royal Commission on Electric Power Planning.

Bridges, J. E. and Preache, M. (1981). Biological influences of power frequency electric fields: A tutorial review from a physical and experimental viewpoint. *Proceedings of the IEEE*, **69**, 1092–1120.

Brush, R. O. and Palmer, J. F. (1979). Measuring the impact of urbanization on scenic quality: Land use change in the Northwest. In G. Elsner and R. Smardon (eds), *Our National Landscape* (p. 55). Washington, DC: USDA Forest Service.

Busby, K., Driscoll, D. and Washbon, W. E. (1974). *A Field Survey of Farmer Experience with 765 kV Transmission Lines*. New York: Agricultural Resources Commission.

Butler, J. C. (1983). *Electric Utility Transmission Lines and Land Use Compatibility: Annotated Bibliography*. Chicago, IL: American Planning Association.

Casper, B. M. and Wellstone, P. D. (1981). *Powerline: The First Battle of America's Energy War*. Amherst: University of Massachusetts.

Crawford, C. O. (1955, July). Appraising damages to land from power line easements. *The Appraisal Journal*, 367–378.

Crowe, S. (1958). *The Landscape of Power*. London: The Architectural Press.

Daneke, G. A., Garcia, M. W. and Delli-Priscoli, J. (eds). (1983). *Public Involvement and Social Impact Assessment*. Boulder, CO: Westview Press.

Delli-Priscoli, J. (1983). The citizen advisory group as an integrative tool in regional water resources planning. In G. A. Daneke, M. W. Garcia, & J. Delli-Priscoli (eds), *Public Involvement and Social Impact Assessment*. Boulder, CO: Westview Press.

DeVore, F. W. (1969). The challenge created by aesthetics. *Proceedings of the American Right of Way Association*. Inglewood, CA, 48–49.

Ducsik, D. W. (1981, April). Citizen participation in power plant siting: Aladdin's lamp or Pandora's box? *Journal of American Planning Association*, 154–166.

Edwards, W. and von Winterfeldt, D. (1987). Public values in risk debates. *Risk Analysis*, **7**(2), 141–158.

Enk, G. A. (1981). *Review of a Transmission Line Visual Assessment Framework* [Report on Phase Two]. Washington, DC: U.S. Department of Energy.

Enk, G., Finin, G. A., Hornick, W. F., Jordan, J. J. and Schneller, K. A. (1980). *Review of a Methodology for Assessing Visual Impacts of Overhead Transmission Lines* [Report on Phase One]. Washington, DC: U.S. Department of Energy.

Ewald, G. A. (1969). Aesthetics and rights of way. *Proceedings of the American Right of Way Association*. Inglewood, CA, 49–51.

Fidell, S., Teffeteller. S. R. and Pearsons, K. S. (1978). *The Effects of Transformer and Transmission Line Noise on People: Community Reaction* (Vol. 3). Bolt, Beranek & Newman.

Fischhoff, B., Slovic, P., Lichtenstein, S., Read, S. and Combs, B. (1978). How safe is safe enough? A psychometric study of attitudes towards technological risks and benefits. *Policy Sciences*, **68**, 64–69.

Fraggalosch, A. (1980). *A Review of B. C. Hydro's Public Participation Process on Transmission Line Planning: A Case Study of the Cheekye-Dunsmuir 500 kV Transmission Line* [Internal working paper]. Vancouver, British Columbia: B. C. Hydro and Environmental Land Use Committee Secretariat.

Fricke, R. N. (1964). Problems encountered in laying out and securing approval of routes for extra high voltage transmission lines. *Proceedings of the American Right of Way Association*. Inglewood, CA, 106–109.

Fridriksson, G., MacFayden, M. and Branch, K. (1982). *Electric Transmission Line Effects on Land Values: A Critical Review of the Literature*. Billings, MT: Mountain West Research, Inc.

Furby, L. (1978). Possessions: Toward a theory of their meaning and function throughout the life cycle. In P. B. Baltes (ed.), *Life-Span Development and Behavior* (Vol. 1). New York: Academic Press.

Furby, L. (1986). Psychology and justice. In R. L. Cohen (ed.), *Justice: A View from the Social Sciences*. New York: Plenum, pp. 153–203.

Furby, L., Slovic, P., Fischhoff, B. and Gregory, R. (in press). The effect of electric power transmission lines on property values, compulsory purchases, and rerouting. *Journal of Environmental Management*.

Gale, M. (1982). *Transmission Line Construction Worker Profile and Community/Corridor Resident Impact Survey*. Billings, MT: Mountain West Research, Inc.

Gatchel, R. J., Baum, A. and Baum, C. S. (1981). *Stress and Symptoms Reporting as Related to the CU Line*. Olney, MD: Human Design Group.

Genereux, J. P. and Genereux, M. (1980). *Perception of Landowners about the Effects of the UPA/CPA Powerline on Human and Animal Health in West Central Minnesota*. St Paul, MN: Minnesota Environmental Quality Board.

Gussow, A. (1977). In the matter of scenic beauty. *Landscape*, **21**(3), 26–35.

Hendrickson, P. L., *et al.* (1974, July). *Measuring the Social Attitudes and Aesthetic and Economic Considerations which Influence Transmission Line Routing*. Seattle, WA: Battelle Northwest Labs.

Jackson, R. H., Hudman, L. E. and England, J. L. (1978). Assessment of the environmental impact of high voltage power transmission lines. *Journal of Environmental Management*, **6**, 153–170.

Jones, G. and Jones, I. (1976). *Visual Impact of High Voltage Transmission Facilities in North Idaho and Northwestern Montana*. Seattle, WA: Jones & Jones.

Katz, M. (1971). Decision-making in the production of power. *Scientific American*, **224**, 191–200.

Kinnard, W. N. (1967, April). Tower lines and residential property values. *The Appraisal Journal*, 35, 269–284.

Kipp, K. W. (1969, June 19). A new role for community planners with electric utility companies. *Public Utilities Fortnightly*, pp. 56–62.

Mason, R. (1982, December). *The Location of Powerlines and Social Conflict*. Paper presented at EPRI Planning Session RP 2069, Palo Alto, CA.

McGuire, W. J. (1969). The nature of attitudes and attitude change. In G. Lindsey & E. Aronson (eds), *The Handbook of Social Psychology* (2nd ed., Vol. 3). Reading, MA: Addison-Wesley, pp. 136–314.

Mitchell, B., *et al.* (1976). *The Long-Term Socio-Economic Impact of an Electrical Power Transmission Corridor on the Rural Environment: Perception and Reality*. Waterloo, Ontario: Royal Commission on Electric Power Planning.

Morgan, M. G., Florig, H. K., Nair, I and Lincoln, D. (1985, February). Power-line fields and human health. *IEEE Spectrum*, 62–68.

Morgan, M. G., Lincoln, D. R., Nair, I. and Florig, H. K. (1983). *An Exploration of Risk Assessment Needs and Opportunities for Possible Health Consequences from 50/60 hz Electromagnetic Fields*. Pittsburgh, PA: Department of Engineering and Public Policy.

Morgan, M. G., Slovic, P., Nair, I., Geisher, D., MacGregor, D., Fischhoff, B., Lincoln, D.

and Florig, K. (1985). Powerline frequency electric and magnetic fields: A pilot study of risk perception. *Risk Analysis*, **5**, 139–149.

Nolfi, J. R. and Haupt, R. C. (1982). *Effects of High Voltage Power Lines on Health: Results from a Systematic Survey of a Population Sample along the 400 kV Pacific Intertie.* Burlington, VT: Associates in Rural Development.

O'Hare, M., Bacow, L. and Sanderson, D. (1983). *Facility Siting and Public Opposition.* New York: Van Nostrand Reinhold Co.

Pohlman, J. C. (1973, April). What *is* the public's opinion on transmission towers and poles? *Electric Light and Power*, 59–61.

Popper, F. J. (1981, April). Siting LULUs. *Planning*, 12–15.

Porter, A. (1980). *Report of the Royal Commission on Electric Power Planning* (Vols. 1–9). Ontario: The Commission.

Priestley, T. (1983a). *Aesthetic Considerations and the Electric Utility Industry: An Introductory Guide to the Literature.* Palo Alto, CA: Electric Power Research Institute.

Priestley, T. (1983b). *Symbolism and Meaning in Environmental Conflict: A Planner's Analysis of the Minnesota Powerline Revolt.* Unpublished manuscript.

Priestley, T. (1983c). *Transmission Lines and Land Use Development.* Edison Electric Institute.

Ray, J. (1978, February 18). The hazards of high wires. *The Nation*, 177–180.

Renn, O. and Swaton, E. (1984). Psychological and sociological approaches to the study of risk perception. *Environmental International*, **10**, 557–575.

Rydant, A. L. (1984). A methodology for successful public involvement. *Social Impact Assessment*, **96–98**, 4–9.

Slovic, P., Fischhoff, B. and Lichtenstein, S. (1982). Why study risk perception? *Risk Analysis*, **2**, 83–93.

Slovic, P., Fischhoff, B. and Lichtenstein, S. (1984). Behavioral decision theory perspectives on risk and safety. *Acta Psychologica*, **56**, 183–203.

Slovic, P., Fischhoff, B. and Lichtenstein, S. (1985). Characterizing perceived risk. In R. W. Kates, C. Hohenemser and J. Kasperson (eds), *Perilous Progress: Managing the Hazards of Technology.* Boulder, CO: Westview.

Slovic, P., Lichtenstein, S. and Fischhoff, B. (1984). Modeling the societal impact of fatal accidents. *Management Science*, **30**, 464–474.

Smith, T. W., Jenkins, J. C., Steinhart, J. S., Briody, K. A. and Schoengold, D. (1977). *Transmission Lines: Environmental and Public Policy Considerations: An Introduction and Annotated Bibliography.* Madison, WI: University of Wisconsin, Institute for Environmental Studies.

Squires, S. (1985, May 20). Beware of electric pollution: evidence suggests rays are harmful. *The Washington Post National Weekly Edition*, 10.

Sweet, D. L. (1970). How are the utilities images to public bodies. *Proceedings of the American Right of Way Association.* Inglewood, CA, 39–47.

Tichenor, P. J., Donohue, G. A. and Olien, C. N. (1980). *Community Conflict and the Press.* Beverly Hills, CA: Sage.

Tyler, T. R. (1984). Justice in the political arena. In R. Folger (ed), *Justice: Emerging Psychological Perspectives.* New York: Plenum.

Vroom, V. R. (1969). Industrial social psychology. In G. Lindzey and A. Aronson (eds), *The Handbook of Social Psychology* (2nd ed., Vol. 5). Reading, MA: Addison-Wesley, pp. 196–268.

Wasserman, H. (1979, August). Revolt of the bolt weevils. *Rolling Stone*, p. 38.

Young, L. B. (1973). *Power over People.* New York: Oxford.

Young, L. B. (1978). Danger: high voltage. *Environment*, **20**, 16–38.

PARENTAL CONCERN ABOUT CHILDREN'S TRAFFIC SAFETY IN RESIDENTIAL NEIGHBORHOODS

TOMMY GÄRLING, ANITA SVENSSON-GÄRLING

Environmental Psychology Research Unit, Department of Psychology, University of Umeå S-901 87 Umeå, Sweden

and JAAN VALSINER

Program in Developmental Psychology, Department of Psychology, University of North Carolina, Davie Hall 013A, Chapel Hill, North Carolina 27514, U.S.A.

Abstract

One hundred and five parents and nonparents responded to a questionnaire consisting of evaluative ratings (general evaluation, social status and safety concern) of six familiar residential neighborhoods; ratings of the traffic accident risk children in the age ranges 2–4, 5–6, 7–9 and 10–12 years run in these neighborhoods; and, finally, ratings of the strengths attributed to factors as causes of traffic accidents (environment, children, parents, drivers and chance). Across neighborhoods and age ranges of children, risk perceptions were found to be related to the rated strengths of the causes. Low-traffic volume neighborhoods were perceived as less risky and were attributed as less strong causes than high-traffic volume neighborhoods were. Perceived risk increased with age of child to a maximum, then decreased. The same relationship with age was found for the attributed causes environment and drivers. The strength of parents as cause was rated to decrease while the strength of child as cause was rated to increase with age. Chance was rated as the weakest cause and the rated strength did not vary across neighborhoods or age. Neither parentship nor gender, singly or in combination, had any clear effects. For parents and nonparents alike the general evaluation of the neighborhoods was influenced by safety concern but not by the particular aspect investigated, i.e. perceived traffic accident risk to children.

Introduction

Technological development has had some well known human costs. One of them is injuries to children caused by motor vehicles in residential neighborhoods. Protective measures have therefore been taken, such as, for instance, the prohibition of motor traffic. Nevertheless, whether these measures are effective, community actions, i.e. the formation of neighborhood groups, letters to the editor and the like, testify to the fact that parents continue to be worried.

Parents' concern and the counter-measures they themselves take in the interest of protecting their children are probably in many cases necessary for the successful prevention of traffic accidents in residential neighborhoods. This is obviously so when the children are very young, but even when they grow older, when they start to explore the outside world, their protection is likely to be mainly the parents' responsiblity. Everything that can be done to facilitate the parents' task should therefore be done, and to this end a better understanding of the problems parents face and how they handle these problems would certainly be of value.

A basic tenet in the present study is that parents act on the basis of their understanding of the potential dangers their children run in residential neighborhoods *vis-à-vis* motor vehicle traffic. Parents may perceive the child, the environment and other people as the major causes of traffic accidents, i.e. as the major threats to the children's traffic safety, and their perceptions in this respect may determine, first, whether they consider it necessary to take any counter-measures, and, second, what counter-measures they should take in such cases.

To illuminate the cognitive processes that account for parents' understanding, a number of questions were raised and subjected to investigation. First, do parents perceive residential neighborhoods to vary in the degree to which they are threatening to the traffic safety of children at different ages? Second, are these perceptions related to attributions of the environment as a cause of traffic accidents, and what particular attributes of the neighorhoods account for that? Third, what is the causal strength of the environment as compared to the strengths of the other perceived causes of traffic accidents? Fourth, if parents perceive the traffic in residential neighborhoods to be unsafe for children, do they therefore evaluate these neighborhoods much lower than they would have done otherwise? In other words, is traffic safety such a salient attribute that it influences parents' evaluations of their residential neighborhoods very negatively, perhaps even though they are perceived as quite satisfactory in other respects? Fifth, and finally, do parents differ from nonparents in their perceptions of children's traffic accident risks, and are their evaluations of residential neighborhoods more negatively affected by these perceived risks than nonparents' evaluations are? Hypotheses related to each one of these questions are derived below.

Causal attribution

Attribution theory in social psychology (Heider, 1958; Jones and Davies, 1965; Kelley, 1972*a*, 1979) has been proposed to account for people's general tendency to attribute causes to their own and other's behavior. Two points seem to be important in relation to parents' attributions of causes of traffic accidents. First, causal schemata, defined as internal representations of the causal structure of the man–environment system, have been found to play a decisive role in causal attribution (Kelley, 1972*b*; Reeder and Brewer, 1979). Second, in causal attribution, personal and environmental factors interact to determine the outcome (Heider, 1958; Hewstone, 1983; Valsiner, 1984).

Causal schemata are acquired from various sources, primary as well as secondary, such as mass media (Wells, 1981). Their function is to make it possible to go beyond perceptual information in order more validly to infer causes of events. It does not follow, however, that causal attribution is invariably veridical. Some events are not deterministic, and, even if they are, knowledge of the causative factors and the circumstances is probably seldom complete. Parents are likely to acquire causal schemata related to children's traffic accidents mainly from secondary sources. Traffic accidents in residential neighborhoods occuring to children are, after all, not frequently encountered. Although near-accidents may be much more frequent, compulsory education and the mass media are probably more important sources of information for parents about traffic accident risks, traffic safety measures and the like. No analysis of this 'educational' process is possible here, but there is no doubt that it is predicated on the general notion that the best tactic in protecting children is to eliminate the causes of traffic accidents. Furthermore, information is conveyed about children's and drivers' ability to handle traffic hazards and about the factors that make this more difficult for them.

As knowledge about causal factors increases, one may come to perceive an event as more deterministic than the case was before. This may pertain to the acquisition of causal schemata related to children's traffic accidents. Nonparents are probably informed about traffic accident risks and the like to almost the same extent as parents are. A greater interest in children's welfare, as well as more previous experience of accidents and near-accidents, may nevertheless lead parents in general to acquire more knowledge than nonparents. However, knowledge differences may also lead to differences in the contents of the causal schemata, i.e. in the relative strengths of the causes attributed to traffic accidents.

Although the main concern in the present study is the attributes of residential neighborhoods that are perceived by parents as threats to the children's traffic safety, that causal attribution is jointly determined by environmental and personal factors must be taken into account. As long as the child lacks many of the skills that the adult possesses, parents are responsible for monitoring the child. If an accident occurs, the cause may therefore primarily be attributed to the parents who may be assumed to have failed in monitoring the child. The environment and the drivers may, however, still be perceived as strong causes since parental monitoring is facilitated if the environment is safe and the drivers cautious. The environment and the drivers may furthermore gain in importance as the child grows older. Parental monitoring becomes less efficient (e.g. when the child starts school), at the same time the child has not yet acquired the skills to manage completely on his/her own. Thus, an unsafe environment and noncautious drivers could be most dangerous at this stage of the child's development. As the child eventually attains the required skills, both the responsibility of the parents and the threats from the environment and drivers, should thereby decrease. Traffic accidents that still occur are possibly then perceived as caused by the child to a greater extent than by other causes. The child may, for instance, be assumed to have taken undue risks.

The hypothesis about the contents of parents' causal schemata, advocated here, leads to the prediction that environmental factors would be perceived as more threatening (i.e. as stronger causes) when the child has reached a certain age (developmental level). Although no exact quantitative prediction is possible, beginning school age may be the critical point. Further predictions that follow are that the strength of parents as a cause would be perceived to decrease with the child's age and the causal strength of the child to increase reciprocally. The perceived causal strength of drivers, finally, may more or less parallel that of the environment. The question should also be raised: what attributes of residential neighborhoods are perceived as dangerous? Although no definite hypothesis can be offered, factors influencing traffic volume and the degree to which children are exposed to motor vehicles (availability of pedestrian crossings, playgrounds, etc.), possibly modified by speed limits and other regulations of the traffic, are likely candidates. Another question concerns whether environmental attributes are perceived to influence traffic accident risks directly, or indirectly, by, for instance, making the task more difficult for drivers. If there are indirect influences, the environmental attributes should not only increase the causal strength attributed to the environment but also that attributed to other causes.

Apart from perceiving traffic accidents as more deterministic, parents may, because they possess more knowledge than nonparents, differ from the latter with respect to the contents of their causal schemata. In particular, the strength attributed to the child as a cause may differ since it is probably in that domain (i.e. knowledge about children's skills) that parents' knowledge is better than nonparents'. This superior knowledge does, of course, primarily pertain to their own children but, to some degree possibly, also to children in general.

Risk perception

The prediction of the occurences of uncertain events, such as children's traffic accidents, is generally held to be a difficult task for lay people and experts alike (Kahneman *et al.*, 1982), and, as Kahneman and Tversky (1973) have argued, there is hardly any other way to make such predictions or judgments than to use heuristics which reduce the judgments to simpler ones. More recently, Tversky and Kahneman (1980) and Kahneman and Tversky (1982) have assumed, first, that causal schemata play a role in lay people's assessments of probabilities, and, second, that a 'simulation' heuristic is used in these assessments. Their notion is that, when faced with the problem of having to assign to an event its probability of occurrence, the possible outcomes (termed 'scenarios') are mentally simulated. The ease with which the event is simulated determines what probability is assigned. Although Kahneman and Tversky (1982) did not explicitly state that causal schemata could be assumed to play their role in probability assessments by making possible the mental stimulation.

To account for how parents perceive children's traffic safety in residential neighborhoods and how they act, the hypotheses are here set out: (i) that parents use their acquired causal schemata to mentally simulate the outcomes of different observed event sequences, (ii) that the risk they perceive is dependent on the outcome of and the properties of the mental simulation and (iii) that the decisions to act and how to act are based on the evaluated risk. Hypotheses (i) and (ii) lead to the general prediction that the perceived traffic accident risk should be related to the perceived strengths of the causes attributed to traffic accidents. For instance, other things being equal, residential neighborhoods perceived to vary in causal strength should be perceived to vary similarly in traffic accident risk. The risk assessments should however also depend on the perceived strengths of other causes, so to predict the risk assessments, the causal strength of the environment and the strengths of the other causes must be combined in some manner. The assumption is that this combination is achieved in the simulation process, and unless more is known about its nature, exact predictions are precluded.

Another general implication is that more accurate causal schemata would lead to more accurate risk assessments. If parents possess more relevant knowledge than nonparents do and their causal schemata for that reason are more accurate, then they would also make more accurate risk assessments. However, since the accuracy of predictions of children's traffic accidents is difficult to assess, the only prediction that could strictly be evaluated is that parents, in accord with the differences in their causal schemata, would make risk assessments that differ from those made by nonparents.

Residential satisfaction

Two different theoretical approaches to residential satisfaction are possible to distinguish (Michelson, 1977, 1980): the more traditional, static, view and an alternative that brings the time dimension into focus. In both these approaches, how residential satisfaction relates to attributes of the neighborhoods in which one lives needs to be modeled. A viable attempt at such modeling has been proposed by Miller *et al.* (1980). According to this model people first evaluate the beliefs they hold about their residential neighborhoods, then they combine these evaluations into an overall evaluation. However, due to a limited ability to take all one's evaluations (or beliefs) into account, a few, very salient, beliefs may come to dominate the overall evaluation at the expense of other less salient beliefs. Saliency may vary across residents in a neighborhood, thus the model should be capable of accounting for individual differences, and it may also vary over time, thus the model is also potentially capable of accounting for some of the dynamic aspects of residential satisfaction.

It has repeatedly been asserted here that parents are inclined to act in the interest of protecting their children from traffic accidents in the residential neighborhood where they live. Traffic safety may consequently stand out as a salient attribute. Nonparents, on the other hand, are not equally responsible for the protection of children, so traffic safety may be less salient to them. If these assumptions about the saliency of traffic safety are combined with the assumption made by Miller *et al.* (1980), the prediction follows that parents' evaluations of their residential neighborhoods should be dominated by the traffic accident risk they perceive the children to run. Nonparents' evaluations should in contrast be less influenced by this concern.

Method

The sets of hypotheses advanced above were subjected to investigation in the present study. Parents and nonparents responded to a questionnaire in which three sets of ratings of six different residential neighborhoods were requested. The neighborhoods were all familiar to the subjects although they, of course, only lived in one of them (which one varied across subjects). The intention was to simulate the situation in which one is faced with the option not to move out of the neighborhood where one lives or to move to another neighborhood located in the same town. Furthermore, the neighborhoods were selected to be representative of the range of traffic hazards to which children in the town are likely to be exposed. They varied in types of dwellings, traffic volume, availability of playgrounds and street patterns.

The sets of ratings consisted of, first, semantic differential scales which were checked for each residential neighborhood. Three sets of scales were used. One set tapped a general/aesthetic evaluative dimension, another set an independent social status dimension and a third set, more directly, tapped safety concern, i.e. the content assumed to dominate parents' evaluations (although care was taken not to contaminate this factor, the criterion variable, with the variables to be used to predict it). Second, subjects rated the traffic accident risk for children in four age ranges (2–4, 5–6, 7–9 and 10–12 years) living in respective neighborhoods. The age ranges were selected to be perceived as referring to fairly homogenous developmental levels of the children and to cover the critical age ranges. Finally, ratings were obtained of the perceived strengths of causes of accidents: children, parents, environment, and drivers. These ratings were obtained under exactly the same conditions as the ratings of the risk. Ratings of the strength of bad luck or chance as 'cause' were also obtained in connection with the ratings of the other causes. The idea was that the ratings of the causal strength of chance would indicate to what degree traffic accidents are perceived as being specifically determined.

Residential neighborhoods

The study was conducted in the town of Umeå located in the northern part of Sweden. In 1983 the town population reached about 83,000 residents after having undergone a rapid change during the 1960s and the 1970s. To meet the housing needs of people moving in, a number of residential neighborhoods have been built around the 19th century town.

Six residential settings (labeled A–F) were selected for the study according to several criteria. The settings should, first, represent the range of traffic hazards to which children living in the town are likely to be exposed. Second, the settings or neighborhoods should be fairly homogenous and still be representative of the districts in which they are located. Third, census tract information and accident statistics should be available. Fourth, in each neighborhood there should be nursery, primary and secondary schools attended by the majority of the children in the age range from 5 to 12 years living there (Note 1).

Neighborhoods A and B were built in the late 1960s. Predominantly multi-family dwellings are located in these neighborhoods. Both have traffic routes separated from pedestrian routes. C Is an older neighborhood with predominantly multifamily dwellings but with no traffic separation and with few playgrounds. The traffic volume is, however, low because the neighborhood is located far from downtown and the major through-traffic arteries. Another neighborhood built in the 1960s, D, is a typical neighborhood located in a quiet district far from downtown, dominated by bungalows, with narrow, often curved, streets and with few playgrounds. Visibility for drivers is generally low in the streets due to fences and greenery. Through-traffic is prohibited in some parts. An older neighborhood is E, which is located in the downtown outside the main business district. The traffic volume is higher but it is in other respects similar to D. Another, similar neighborhood, F, is partly located in the business district where there is heavy through-traffic of buses, private cars, and cabs.

The number of children in the age ranges of 2–4, 5–6, 7–9 and 10–12 years, respectively, living in each of the residential neighborhoods A–F is given in Figure 1. The figures are from the 1982 census. The most recent available accident statistics (from 1976 to 1980) are also presented. The number of children per thousand, living in the neighborhood, reported to the police to have been injured in a traffic accident is given for each age range and neighborhood. The police-reported accidents are few, which makes the frequencies unreliable. Nevertheless, they suggest that the settings located in downtown are more risky, and that the risk increases with the children's age.

In the rating procedure, the residential neighborhoods were labeled by their commonly used names. Additional information was provided by means of maps shown to the subjects in a manner which will be described below.

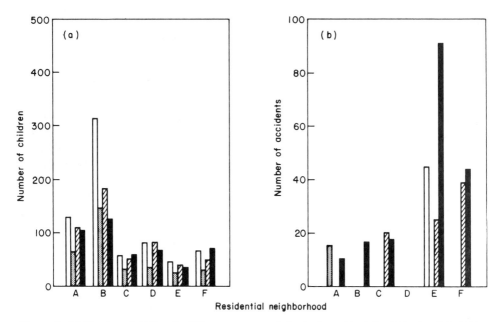

FIGURE 1. (a) Number of children in different age ranges living in the residential neighborhoods in 1982, and (b) number of police-reported traffic accidents to children in the different age ranges per thousand children living in each residential neighborhood. Children's ages: □, 2–4 years; □, 5–6 years; ▨, 7–9 years; ■, 10–12 years.

Subjects

One-hundred and eleven students attending a community college volunteered as subjects in return for payment. Complete data were obtained for 30 male and 45 female parents and 12 male and 18 female nonparents. The subjects had lived in one or several of the residential neighborhoods in the town for from less than a year to all their lives ($M = 12 \cdot 2$ and $8 \cdot 9$ years, respectively, for parents and nonparents). The parents' ages ranged from 18 to 54 years ($M = 35 \cdot 1$ years), the nonparents' from 18 to 38 years ($M = 23 \cdot 4$ years). Twenty seven of the parents had one child (mean age $7 \cdot 6$ years), 30 two children (mean ages of first and second child $13 \cdot 4$ and $10 \cdot 2$ years, respectively), and 18 three children (mean ages $16 \cdot 5$, $13 \cdot 3$ and $9 \cdot 1$ years). The majority of the subjects (83) were drivers. Another 11 subjects had a driving licence but did not drive regularly.

Procedure

The subjects were approached through their teachers in the community college. During regular class hours, the teachers asked the students to participate in the study. The experimenter administered a booklet containing all the questions, the maps and the rating scales. No student refused to participate but six had to be discarded because their data were found to be incomplete. Each session took about one hour.

Before being asked to participate, the students had been given general written information about the study handed out by the teachers a few days in advance.

Each session then started with the experimenter reviewing this information. Any questions at this time or later were answered by paraphrasing appropriate parts of the instructions. After the general information had been given, the subjects worked through the first part of the booklet containing questions about gender, age, education, occupation, marital status, number, age and gender of children, and driving experience and places of residence while living in the town. The subjects responded anonymously.

The second part of the questionnaire consisted of the semantic differential scales by means of which each of the residential neighborhoods were to be rated. The subjects could not be expected to be equally familiar with all the neighborhoods. Therefore, on the pages in the booklet preceding the page with the semantic differential scales, a map of Umeå indicating the general location of the neighborhood was first shown. Then, a detailed map indicating the exact location, including a drawn border demarcating the neighborhood from its surroundings, was shown. While looking at the overall map, the subjects were asked to rate how familiar the neighborhood was to them. Beneath the map, on the same page, a 100 mm long, horizontal graphic scale with unit markers every 10 mm was provided. The endpoints were defined as 'know very much' (right end-point) and 'know very little' (left end-point), respectively. A cross was to be drawn anywhere on the line indicating the degree of familiarity with the neighborhood.

The set of rating scales on the following page consisted of nine semantic differential type of bipolar scales. Graphical scales, 100 mm long, with unit markers every 10 mm were also used. The end-points were from right to left defined as Good–Bad, Attractive–Unattractive, Beautiful–Ugly, High–Low Status, Exclusive–Inexclusive, Expensive–Cheap, Safe–Unsafe, Peaceful–Not Peaceful and Dangerous–Not Dangerous. The first three scales have previously been found to tap a general/aesthetic evaluative dimension (Gärling, 1976*a*,*b*, 1980), the following three scales an independent social status dimension and the last three were new, included, as already mentioned, because they were hypothesized to be more directly related to the content assumed to dominate the parents' evaluations of the neighborhoods. The subjects checked the scales in their order of appearance from top to bottom on the page in the booklet. This order was determined randomly. The order in which the neighborhoods were rated were individually randomized. The subjects were asked to report their own personal opinion, trying to ignore what they might think general opinion is. The subjects were not told anything about the expected relationship between the evaluations of the neighborhoods and the perceived risk of children's traffic accidents.

In the third part of the questionnaire the detailed maps of the neighborhoods were shown once again in another random sequence. This time the task was to judge the risk the subjects believed children in different age ranges (2–4, 5–6, 7–9 and 10–12 years) run of having a traffic accident. A graphic scale was provided for each age range on a single page succeeding the map. The end-points were defined from right to left as 'very great risk' and 'very small risk' respectively. The instructions emphasized that the risk should be judged for one child living in the particular neighborhood, ignoring differences due to the base rate of children living there, as well as differences due to gender and similar, possibly relevant characteristics.

The detailed maps were finally shown in still another random sequence in the fourth part of the questionnaire. The task now was to rate the strength with which the subjects attributed the following factors as a cause of traffic accidents to children (in order of appearance from top to bottom): the environment, the children, the parents, the driver and chance. The end-points of the graphic scales were from right to left defined as 'very much strength' and 'very little strength', respectively. It was emphasized that the ratings should apply to the causes of a traffic accident in the particular neighborhood which had occurred to a child in the specified age range. The subjects were also told that traffic accidents may be difficult to explain, and that there are many factors which alone or in combination may account for their occurrence.

Scoring and preliminary analyses

Raw scores were obtained, for each subject and scale, by measuring the distance in cm from the left end-points. These scale values, ranging maximally from zero to ten, were subjected to statistical analyses as will be described below. No corrections for differences in familiarity with the neighborhoods seemed called for in these analyses, although reliable differences were observed. An analysis of variance (ANOVA) showed that the neighborhoods differed reliably with regard to rated familiarity, $F(5, 505) = 12 \cdot 26$, $P < 0 \cdot 001$. Neighborhoods B and F were rated as reliably more familiar than the other neighborhoods, C as less familiar ($P = 0 \cdot 001$ [Note 2]). There was also a reliable difference between men and women ($M = 5 \cdot 12$ and $4 \cdot 20$, respectively), $F(1, 101) = 8 \cdot 38$, $P < 0 \cdot 01$, and a weak but reliable interaction between parentship, gender, and neighborhood, $F(5, 505) = 2 \cdot 96$, $P < 0 \cdot 05$. The results of analyses of covariance (CANOVAs) showed, however, that the familiarity ratings had no reliable linear relationships with any of the other rating variables (the ratings of the causes, the risk, and the ratings on the semantic differential scales, respectively) ($P = 0 \cdot 05$).

Results

Causal attribution

Figure 2 shows the means of the ratings of the strength of each cause of traffic accidents to children, living in the residential neighborhoods A to F, plotted against age range of child. These plots give general support to the hypothesis about the content of parents' causal schemata related to children's traffic accidents. First, the causal schemata may be considered to be deterministic because chance is on average rated as a less strong cause than any of the other causes. Second, the rated causal strength of parents decreases and, at a somewhat faster rate, the rated strength of child as a cause increases with the age of child. Furthermore, the rated causal strengths of driver and environment are related to age by inverted U-shaped functions, more clearly in the latter than in the former case. The only pronounced change across the neighborhoods, finally, is that the rated causal strength of the environment varies. An ANOVA (subjects parentship × gender) × cause × age of child × residential neighborhood) yielded highly significant main effects of cause, $F(4, 400) = 22 \cdot 03$, $P < 0 \cdot 001$, of age of child, $F(3, 303) = 14 \cdot 9$, $P < 0 \cdot 001$, and of neighborhood, $F(5, 505) = 17 \cdot 01$, $P < 0 \cdot 001$, respectively. The two-way interactions cause × age and cause × neighborhood, respectively, also reached significance, $F(12, 1212) = 51 \cdot 83$, $P < 0 \cdot 001$, and $F(20, 2020) = 9 \cdot 33$, $P < 0 \cdot 001$. Parentship and gender, however, gave rise to few

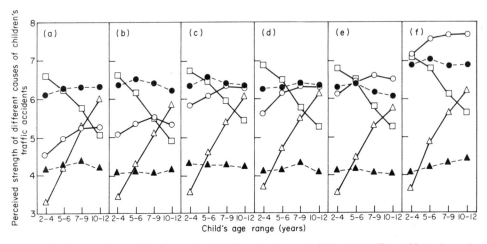

FIGURE 2. Mean ratings of the strengths of different causes of children's traffic accidents in each residential neighborhood as functions of the age ranges of child. (a)–(f), Neighborhoods A–F. ○, Environment; △, children; □, parents; ●, driver; ▲, chance.

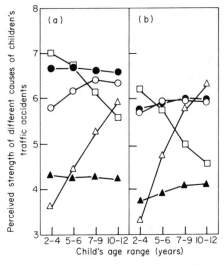

FIGURE 3. Parent's (a) and nonparent's (b) ratings of the strengths of different causes of children's traffic accidents as functions of the age ranges of the child. ○, Environment; △, children; □, parents; ●, driver; ▲, chance.

significant effects, indicating that individual differences in the causal schemata are at best slightly accounted for by these factors singly or in combination. Parents rated the causes as stronger on average than nonparents did, but this level difference was not significant ($P < 0.25$). There was furthermore a tendency to an interaction between parentship and cause ($P < 0.25$), a significant interaction between parentship, cause and age of child, $F(4, 404) = 1.89$, $P < 0.05$, and a significant interaction between parentship and neighborhood, $F(5, 505) = 2.56$, $P < 0.05$. Gender was not involved in any significant effects ($P = 0.05$).

As already noted in support of the hypothesis, chance was rated as a less strong cause to children's accidents than the other causes. This difference did not quite reach significance in the subsequent tests ($P < 0.05$), but, as Figure 2 shows, the only consistent exception is that child is rated as a less strong cause for the age ranges 2–4 years. A separate ANOVA on the ratings of the causal strength of chance showed that the ratings did not vary reliably with age of child, nor across neighborhoods. Parents did not differ reliably from nonparents. A significant interaction between parentship and age of child was however found. Figure 3 shows that the non-parents rated the causal strength of chance to increase slightly with age of child, whereas the parents' ratings display no such increase.

The rated strength of parents as attributed cause decreased with age of child while the rated causal strength of child increased. Separate ANOVAs indicated that the main effects of age of child were significant, and trend analyses yielded significant linear trends associated with age. As Figure 2 suggests, there tend to be cubic deviations from the linear trend for parents, and quadratic (negatively accelerated) deviations for children ($P < 0.01$ in both cases). Parentship yielded no significant effects. The plots in Figure 3 suggest, however, that parents, as compared with nonparents, rate the child as a weaker cause and to increase less fast with age. A less clear, reversed tendency applies to the ratings of parents as cause.

Drivers and environment tended on average to be rated as the strongest causes ($P < 0.25$). Figure 2 shows that the only consistent exceptions were that parents are rated as a stronger cause for the age ranges 2–4 years, and that, in neighbor-hoods A and B, parents are rated as a stronger cause than environment. Separate ANOVAs yielded a reliable main effect of age of child for environment as attributed cause, but not for drivers. Trend analyses indicated that, in the former case, the linear and quadratic trends both approached significance ($P < 0.05$ and $P < 0.01$, respectively), in the latter case only the quadratic trend tended to be significant ($P < 0.01$). For environment as attributed cause, the maxima tended to occur for higher age ranges (7–9 and 10–12 years) rather than for drivers. No significant effects were associated with the factor parentship.

Finally, the most important difference between the residential neighborhoods was that the causal strength attributed to environment varied. This effect was significant in an ANOVA on the ratings of the causal strength of environment. The traffic-separated neighborhoods A and B differed reliably from the low-traffic volume neighborhoods C, and D and E which in turn differed reliably from neighborhood F, located in the downtown business district. The causal strengths attributed to parents and drivers also varied reliably across neighborhoods. However, none of the mean differences were large enough to be reliable.

The significant and nearly significant interactions involving parentship obtained in the overall ANOVA were, as has been reported, mainly accounted for by tendencies in subsequent tests of significance. Most notable was that the parents, as compared with the nonparents, rated children as a less strong cause of traffic accidents to children, and that they rated the causal strength of children to increase less quickly with the age of child.

Risk perception

The means of the ratings of the risk are plotted in Figure 4. The hypothesized

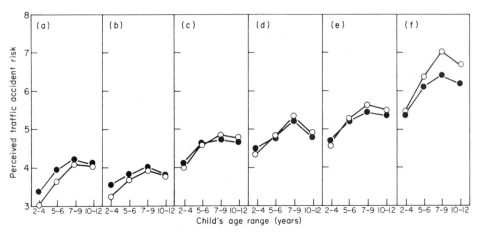

FIGURE 4. Mean rated risk of children's traffic accidents in each residential neighborhood as functions of the age ranges of the child (○) and means adjusted for a linear combination of the ratings of the strengths of the causes used as covariates in the analysis (●). (a)–(f), Neighborhoods A–F.

correspondence with the ratings of the strengths of the causes of children's traffic accidents is indicated. As would be predicted from a weighted linear combination of the ratings of causal strengths i.e.

$$Y_R = A + B_E X_E + B_C X_C + B_P X_P + B_D X_D + B_{CH} X_{CH}$$

(where Y_R denotes rated risk; X_E, X_C, X_P, X_D and X_{CH} rated strengths of the causes environment, parent, child, drivers and chance, respectively; and the Bs are weight coefficients), the perceived risk increases with age of child up to the age ranges of 7–9 years, then decreases. An ANOVA yielded a highly significant effect of age of child, $F(3, 305) = 21.55$, $P < 0.001$, due to both linear and quadratic trends ($P < 0.001$ in both cases). Furthermore, as would be predicted, the perceived risk varied reliably across residential neighborhoods, $F(5, 505) = 44.64$, $P < 0.001$. The risk was rated as reliably higher in neighborhood F, located in the downtown business district, than in the low-traffic volume neighborhoods C, D and E which in turn were rated as more risky than the traffic-separated neighborhoods A and B. As substantiated by a weak but reliable interaction between age of child and neighborhood, $F(15, 1515) = 2.22$, $P < 0.01$, the risk tended also to be rated to increase somewhat faster with age of child in the most risky neighborhoods D, E and F than in the other neighborhoods. The individual differences in risk perception did not seem to be accounted for by parentship and gender. No effects involving these factors, singly or in combination, approached significance ($P = 0.05$).

A direct test of the correspondence between, on the one hand, the ratings of risk and, on the other, the linear combination of the ratings of the strengths of the causes specified by the equation above, was accomplished by means of multiple analysis of covariance (MCANOVA) (Kirk, 1982). In this analysis the ratings of causal strengths were covariates, the ratings of risk dependent variable. If the effects of the independent variables (age of child and neighborhood) are accounted for by the linear combination of the ratings of the causes, then these effects should not remain significant in the MCANOVA. The variation across age of child and across

TABLE 1

The linear regression equations which in the MCANOVA were used to predict the ratings of risk from the ratings of the strengths of causes[a]

Neighborhood
$Y_R = 0.26\ X_E + 0.18\ X_C + 0.13\ X_P - 0.12\ X_D - 0.17\ X_{CH} + 2.89$ Multiple correlation = 0.659, $F(5,500) = 17.96$, $P < 0.001$
Age of child
$Y_R = 0.17\ X_E + 0.28\ X_C + 0.17\ X_P + 0.01\ X_D - 0.07\ X_{CH} + 1.44$ Multiple correlation = 0.667, $F(5,298) = 13.52$, $P < 0.001$
Neighborhood × age of child
$Y_R = 0.08\ X_E + 0.03\ X_C + 0.02\ X_P - 0.02\ X_D + 0.07\ X_{CH} + 3.74$ Multiple correlation = 0.602, $F(5,1510) = 3.58$, $P < 0.01$

[a]The denotations of causes are as follows: E, environment; C, children; P, parents; D, drivers; CH, chance.

neighborhood, respectively, were reduced but not eliminated by the covariate adjustment (see Figure 4). For age, F was reduced to 5.62 ($P < 0.001$ for 3 and 298 df); F for neighbourhood was reduced to 23.4 ($P < 0.001$, 5 and 500 df) and F for the interaction effect marginally reduced to 2.19 ($P < 0.01$, 15 and 1510 df). The MCANOVA also made it possible to estimate the parameters of the equation separately for age of child and neighborhood, and for their interaction, as well as the multiple correlation coefficients between the risk ratings and the ratings of the strengths of the causes. Table 1 shows that there are significant relationships between the dependent variable and the covariates. The causes environment and child account for the major part of the variation in the risk ratings across all three sources, although the relative weights differ somewhat. The weight for environment is larger for the variation across neighborhoods and for the interaction, the weight for child larger for the variation across age of child.

Residential neighborhood attitudes

The ratings on the semantic differential scales were averaged for the sets of three scales that measured general evaluation, social status, and safety concern (Note 3) respectively. Figure 5 shows the means across subjects plotted for each residential neighborhood. The risk ratings, averaged over the four age ranges of children, are also plotted in the figure. As can be seen, the three evaluative measures follow similar trends across neighborhoods. However, an overall ANOVA yielded a significant interaction between scales and neighborhoods, $F(10, 1010) = 25.23$, $P < 0.001$. The main effect of neighborhood also reached significance ($F(5, 505) = 61.00$, $P < 0.001$) but neither of the factors, parentship or gender, were involved in any significant effects ($P = 0.05$). Separate ANOVAs showed that the effect of neighborhood was significant on each measure: $F(5, 505) = 54.77$, $P < 0.001$, $F = 46.04$, $P < 0.001$ and $F = 45.37$, $P < 0.001$, for general evaluation, social status and safety concern, respectively. Neighborhoods C, D and È, rated low or intermediate on risk, received high evaluations on all three measures. A, Low on risk, was intermediate on the evaluative measures, and B was low on risk and the evaluative measures.

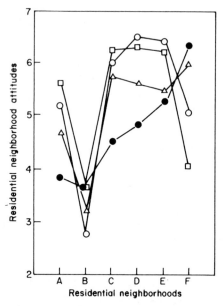

FIGURE 5. Mean semantic differential scale ratings of each residential neighborhood with respect to general evaluation (○), social status (△), and safety concern (□), and mean ratings of risk (●) of children's traffic accidents in each residential neighborhood.

Neighborhood F, the neighborhood rated as most risky, was rated intermediate on general evaluation, high on social status and low on safety concern.

In disagreement with the hypothesis, the results did not show, as has been pointed out already, that there were any reliable differences between parents' and nonparents' evaluations of the neighborhoods. The question may still be raised whether, for all subjects considered as a group, the general evaluation was related to the ratings of traffic accident risks to children. A CANOVA on general evaluation with the mean risk ratings as covariate yielded a slightly negative relationship ($r = -0.199$). However, although significant; $F(1, 504) = 11.95$, $P = 0.01$, the covariate had marginal adjustment effects (F for neighborhood was 57.88). The relationship between safety concern and the risk ratings was somewhat higher ($r = -0.356$, $F(1, 504) = 36.05$, $P < 0.001$) but the covariate adjustment still marginal ($F = 49.78$ for neighborhood). According to a more complex hypothesis, the general evaluation may be a linear combination of social status, safety concern, and traffic accident risk. In support, a MCANOVA yielded a highly significant multiple correlation for this prediction equation ($R = 0.966$, $F(3, 302) = 207.92$, $P < 0.001$). The covariate adjustment was furthermore substantial ($F = 9.04$ as compared to 54.77 in the ANOVA). Social status ($B = 0.35$) and safety concern ($B = 0.63$) contributed however much more to the prediction than did the risk ratings ($B = 0.02$).

Discussion

The present results lend support to the hypothesis about the content of parents' causal schemata related to children's traffic accidents in residential neighborhoods. Furthermore, support was obtained for the hypothesis that the traffic accident risk as perceived by parents depended on the content of their causal schemata, i.e. the strengths attributed to causal factors. No direct observations of the process of risk assessments were made but the demonstrated statistical relationship is nevertheless indicative of the hpothesised properties of this process. Differences had also been predicted between parents and nonparents on the basis of the assumption that parents possess more relevant prior knowledge than nonparents do. Only slight evidence for such differences was however found. This also pertained to the hypothesis that parents, when evaluating the neighborhoods, would be more negatively influenced by their perceptions of children's traffic accident risks. Although safety concern (and social status) appears to have been important for the evaluations of the neighborhoods, there was only weak support for the notion that traffic accident risks were salient, and, at any rate, that it was more salient for parents than for nonparents.

In addition to providing support, the present results may also be taken as a basis for further specification of the hypothesis about the content of parents' causal schemata. That the driver and the environment were perceived as stronger causes on average than parents and child, and that the causal strength of child was perceived to increase faster with the age of the child than the causal strength of parents decreased, were aspects of the results not actually predicted. These aspects need to be encompassed by the hypothesis. That the strength of child, as attributed cause, increased according to a negatively accelerated function of age of child, and the result that the causal strength attributed to parents decreased according to a function containing a cubic component were other aspects of the results which the hypothesis did not predict. They seem, however, plausible if it is assumed that the strengths of these causes reach asymptotic levels as the child grows older.

The risk ratings were found to be statistically associated with the ratings of the strengths of the causes. The relationship was however not strong enough to account for all the variance in the ratings due to neighborhood and age of child. That subjects were unable to use the rating scales in a way which they were supposed to do is a possible artefactual explanation. In particular, subjects may have failed in using commensurate scale units when rating risk as compared to when rating the strengths of the causes. Another possibility might be that nonlinear models would do better in predicting the risk ratings, but the residuals in the prediction from the linear combination did not suggest that. Methods not relying on quantitative ratings made by subjects but still allowing quantitative modeling of relationships may be needed in further studies.

The variance in perceived risk across the residential neighborhoods was found to depend mainly on the strength attributed to the environment as a cause. That both the perceived risk and the causal attributions may correspond to the objective risk is, furthermore, suggested by the accident statistics that were presented in Figure 1. The perceived (and objective) risk appears to increase with increased traffic volume in the neighborhoods but further research is clearly needed to specify the attributes of neighbourhoods that are perceived as threatening to the children's traffic safety.

Both perceived risk and the causal strengths attributed to the environment, and the drivers were found to increase with age of child to a maximum, then to decrease. Whether this corresponds to the objective risk cannot be told for sure but it does not seem unlikely that the objective risk actually increases monotonically with age (cf. Figure 1). If so, adults may make a possibly serious error when judging traffic accident risks to children. This may call for the implementation of some method of judgmental correction (Kahneman *et al.*, 1982). This finding and the findings that the causal strength attributed to the child increased and that the causal strength attributed to parents decreased with the age of children are at any rate in agreement with the predictions of the content of the causal schemata. Somewhat speculatively, the child's ability to master pedestrian activity on her/his own may be a primary judgment underlying the causal attributions. Knowledge of children's development is possibly a prerequisite for such judgments. However, an assessment of the demands imposed by the environment and the behavior of drivers also needs to be taken into account (Heider, 1958). The need for parental monitoring is probably evaluated on the basis of the primary judgments of the child's ability. A problem arises at a certain age of the child when a need for parental monitoring is felt at the same time as circumstances may make this difficult. This may happen because the child starts school but perhaps also because children themselves oppose monitoring. The environment and the drivers then become particularly threatening.

The causal schemata appeared to be mainly deterministic, since chance was rated as the weakest cause. No difference was found in this respect between parents and nonparents. Such differences had been hypothesised on the basis that nonparents had acquired less knowledge about children's traffic accidents and related issues and that less knowledge of a domain leads to less deterministic causal models. Both these premises may, however, be false. Lack of knowledge may not preclude the existence of causal models, only the existence of accurate models. Furthermore, most people, with a minimum of education, living in a highly technological society may, in fact, possess the kind of knowledge comprising the causal schemata pertaining to children's traffic accidents presently under investigation. It should also be noted that some differences between parents and nonparents were observed. Parents attributed less strength to child as a cause than nonparents did, and, parents, in contrast to nonparents, rated the causal strength attributed to child as increasing less quickly with the age of the child. This finding may reflect differences in knowledge about children's development which is precisely the domain of knowledge where differences would be expected.

When they were asked to evaluate the neighborhoods neither parents nor nonparents appeared to be much influenced by their perceptions of children's traffic accident risks. The results nevertheless suggested that safety concern may be salient. At least two factors may account for the negative findings. First, perhaps less obviously, the kind of rating task the subjects performed may have tapped a more 'detached' (cognitive) evaluation of the neighborhoods than would have been the case if subjects had been faced with a real-life decision alternative (e.g. whether to move in with their families). However, it is not clear whether a cognitive evaluation, in contrast to a more affectively toned one, is less susceptible to the kind of selection process that Miller *et al.* (1980) postulate. A second, more plausible explanation, is that children's traffic accident risk, even though it was rated to vary across the neighborhoods, is not high enough to become salient. Traffic accidents to children are, after all, rare events, and much has been done to prevent their occurrence. This does not of course exclude the possibility that they would become salient if, for instance, certain changes, such as the construction of a highway in the vicinity, were proposed (Kahneman *et al.*, 1982).

Other safety concerns, furthermore, seemed to dominate the general evaluation of the neighborhoods. Many parents had children who were teenagers. It is understandable then if other concerns were more salient than traffic accident risk. The neighborhood which received the lowest evaluation was also rated as the least risky. However, it has a reputation of being somewhat socially degraded. Parents, particularly if their children are teenagers, may be more worried for their children in such a neighborhood than if traffic safety is low. Both parents and nonparents may also be worried for their own sake, so in this case there is no real conflict of interests.

Notes

(1) In Sweden all children attend nursery school at the age of 5 or 6 years, primary school between 7 and 9 years, and secondary school between 10 and 12 years. Children living in a community such as Umeå usually go to schools located in their residential neighborhoods.

(2) The overall ANOVAs, the CANOVAs, and the MCANOVAs used the conventional significance levels ($P = 0.05$, $P = 0.01$ and $P = 0.001$) but, since a very large number of statistical tests was carried out, all the subsequent analyses used only the strictest significance level. If an analysis was followed by *post hoc* tests, the Scheffé method was used.

(3) The measure safety concern was labeled in the reverse direction to the scales labelling (e.g. safe as the positive pole). In the text and in the figure, the direction of the measure referred to is congruent with the labeling of the scales.

Acknowledgments

The study was financially supported by grant No. F 902/82 to the first two authors from the Swedish Council for Research in the Humanities and the Social Sciences. A shortened version of the article was presented at the 8th International Conference on 'Environment and Human Action', Berlin (West), 1984, and at the Inaugural European Conference on Developmental Psychology, Groningen, The Netherlands, 1984.

References

Gärling, T. (1976a). A multidimensional scaling and semantic differential technique study of the perception of environmental settings. *Scandinavian Journal of Psychology*, **17**, 323–332.

Gärling, T. (1976b). The structural analysis of environmental perception and cognition. *Environment and Behavior*, **8**, 385–415.

Gärling, T. (1980). A comparison of multidimensional scaling with the semantic differential technique as methods for structural analysis of environmental perception and cognition. *Umeå Psychological Reports*, No. 155.

Heider, F. (1958). *Psychology of Interpersonal Relations*. New York: Wiley.

Hewstone, M. (ed.) (1983). *Attribution Theory*. Oxford: Basil Blackwell.

Jones, E. E. and Davies, K. E. (1965). From acts to dispositions: The attribution process in person perception. *Advances in Experimental Social Psychology*, **2**, 219–266.

Kahneman, D., Slovic, P. and Tversky, A. (eds.) (1982). *Judgment under Uncertainty: Heuristics and Biases*. Cambridge: Cambridge University Press.

Kahneman, D. and Tversky, A. (1973). On the psychology of prediction. *Psychological Review*, **80**, 237–251

Kahneman, D. and Tversky, A. (1982). The simulation heuristic. In D. Kahneman, P.

Slovic and A. Tversky (eds.), *Judgment under Uncertainty: Heuristics and Biases.* Cambridge: Cambridge University Press, pp. 201–208.

Kelley, H. H. (1972a). Attribution in social interaction. In E. E. Jones, D. E. Kanouse, H. H. Kelley, R. E. Nisbett, S. Valins and B. Weimar (eds.), *Attribution: Perceiving the Causes of Behavior.* Morristown, New Jersey: General Learning Press, pp. 1–26.

Kelley, H. H. (1972b). Causal schemata and the attribution process. In E. E. Jones, D. E. Kanouse, H. H. Kelley, R. E. Nisbett, S. Valins and B. Weimar (eds.), *Attribution: Perceiving the Causes of Behavior.* Morristown, New Jersey: General Learning Press, pp. 151–174.

Kelley, H. H. (1979). *Personal Relationships: Their Structures and Processes.* Hillsdale, New Jersey: Erlbaum.

Kirk, R. E. (1982). *Experimental Design,* 2nd Edit. Monterey, California: Brooks/Cole.

Michelson, W. (1977). *Environmental Choice, Human Behavior, and Residential Satisfaction.* New York: Oxford University Press.

Michelson, W. (1980). Long and short range criteria for housing choice and environmental behavior. *Journal of Social Issues,* **36**, 135–149.

Miller, F. D., Tsembris, S., Malia, G. P. and Grega, D. (1980). Neighborhood satisfaction among urban dwellers. *Journal of Social Issues,* **36**, 101–117.

Reeder, G. D. and Brewer, M. B. (1979). A schematic model of dispositional attribution in interpersonal perception. *Psychological Review,* **86**, 61–79.

Tversky, A. and Kahneman, D. (1980). Causal schemata in judgments under uncertainty. In M. Fishbein (ed.), *Progress in Social Psychology* (Volume 1), pp. 117–128. Hillsdale, New Jersey, Erlbaum.

Valsiner, J. (1984). Conceptualizing intelligence: From an internal static attribution to the study of the process structure of organism-environment relationships. *International Journal of Psychology,* (in press).

Wells, G. L. (1981). Lay analyses of causal forces on behavior. In J.H. Harvey (ed.), *Cognition, Social Behavior, and The Environment.* Hillsdale, New Jersey: Erlbaum, pp. 309–324.

PREDICTION OF ENVIRONMENTAL RISK OVER VERY LONG TIME SCALES

JOHN C. BAIRD

Departments of Psychology and Mathematics in the Social Sciences, Dartmouth College, Hanover, New Hampshire 03755, U.S.A.

Abstract

Undergraduate students ($N = 128$) drew curves to reflect the perceived risk over time for nine environmental factors (population, food, water, environmental pollution, disease, genetic defect, mental disorder, nuclear power and nuclear war). Each curve began half-way up the y axis at the year 1982. Four upper time bounds (year) were employed: 2000, 3000, 4000 and 7000. Average statistical profiles were then developed to represent entire sets of curves. Subjects were most optimistic about disease factors and most pessimistic about nuclear power and nuclear war. The upper time limit seemed to have little effect on the results; these students' view of the future appears to be the same over relatively short (20 years) and very long (5,000 years) time scales.

Introduction

Mental models of the likelihood of future events may well be important in shaping the attitudes and behavior of people in both the short- and long-term. As such, the investigation of perceived environmental risk is of interest to both applied and theoretical psychology.

An individual's view of environmental risk can be directed toward the past, the present or the future. Most studies of perceived risk request judgments in the context of the present or in a hypothetical context in the near future (e.g. over the next ten years). By 'present risk' we also imply a future time, since to be at risk means to anticipate a future event. So after all, the assessment of environmental risk involves predictions of future events, their probability and possible impact on an individual or group. In all such cases one assumes that judgments are grounded in the person's past experience with the risk factor under consideration; either through direct personal experience or through the reports of others (e.g. TV, radio, newspapers, books, conversations).

For a number of years, human geographers have studied the perception of environmental risk for natural disasters such as floods, hurricanes and tornadoes (reviewed by Kates, 1976). In particular, they find that prediction of the frequency of future floods is linked with the relative frequency of floods experienced in the past, and that there is a definite tendency to underestimate the potential danger regardless of one's previous experience. According to Kates (1976) a person's most recent experience dominates these predictions of the future, but negative experiences fade rather quickly from memory. One does not expect the future to be quite as bad as the past.

More recently, Slovic *et al.*, (1982) have reviewed a set of psychological experiments comparing the assessment of risk with actual frequency data from the past in regard to a wide variety of risks, ranging from crime and motor vehicle accidents to the failure of one's car to start. They report that people show large errors of judgment, but that whether one underestimates or overestimates the frequency of a hazard depends on the availability of relevant information (often times by the amount of media coverage), as well as on the exact format (context) in which predictions are given. For example, the news media are inclined to report far more deaths due to homicide than due to cancer, and the public tends to grossly overestimate the frequency of the former in respect to the latter.

The research just noted is characterized by three major features. (1) There are hard data of some sort against which a subject's risk assessments can be compared. (2) A point estimate is obtained; that is, a single numerical estimate is given to represent perception of risk at a fixed point or interval of time (e.g., the present or over the next x years). (3) Future time frames are relatively short (usually not exceeding ten years).

The experiment reported here differs from those done previously in that there are no objective data against which to compare risk estimates, the estimates are represented on a continuum stretching over time, and the future time frames are very large (hundreds and thousands of years). In order to obtain data under these conditions a new method of securing risk judgments is introduced, and the results are treated in entirely novel ways.

Method

Subjects

The participants were 128 Dartmouth undergraduates enrolled in an introductory perception course. These students are primarily from upper middle class backgrounds and are chiefly oriented toward professional careers (law, business and medicine). The age, career goals and nationality of the subjects lead one to expect they would be more future oriented than other groups (DeVolder, 1979; Kluckhohn and Strodtbeck, 1961; Svenson, 1985). There were an approximately equal number of males and females.

Procedure

Subjects predicted the amount of risk associated with nine different issues of global concern: population, food, water, environmental pollution, disease, genetic defect, mental disorder, nuclear power and nuclear war. Each potential hazard was identified with a graph depicting the x and y axes but otherwise blank. The y axis was always labeled Risk level, the x axis was labeled Year. The units on the x axis always begun with the year 1982 and increased in roughly equal time intervals either to the year 2000, 3000, 4000 or 7000. Different groups of subjects ($N = 16$) received different upper time limits, as specified by the years noted above. One-half of the subjects ($N = 64$) predicted risk for six of the items, the other half ($N = 64$) predicted risk for the remaining three. The blank graphs were all the same size and were presented together on a standard piece of typing paper.

Sample instructions were as follows: 'Below are six graphs; each represents a potential human risk in the years to come. Indicate your predictions (increases, decreases, whatever) about each risk by drawing a function on each graph starting with the current value (open circle) given on the y axis (half-way up) and continuing over the period of time given on the x axis. For example, if you think the human risk from a factor will increase over time your function should increase, etc. A single function may go up and down or stop at any point'.

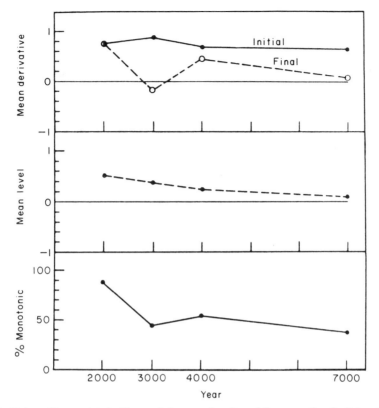

FIGURE 1. Risk profile associated with population as a function of the upper time limit (year): (bottom) the percent of monotonic curves; (middle) the mean level at the final point of the curve in respect to the current level (initial point); (top) initial and final derivatives of the curves (mean values). Data are based on 16 curves at each time limit. For more details see the text.

Results

The most straightforward analysis would simply be to average the height of the drawn curves for each item separately. This does not seem desirable, however, because the issues are so general and the future time scales so unfamiliar. Rather, it would seem better to obtain a representation of each set of curves that captures their general characteristics, without concern for the exact quantitative value of perceived risk at each point in time. Toward this end, five indices were developed to characterize each curve to serve as the risk prediction for each item. These indices can be labeled briefly as follows: (1) monotonicity, (2) relative level, (3) initial derivative, (4) final derivative and (5) termination point. Determination of all indices was based on visual inspection of the curves, with the aid of a straight edge.

(1) *Monotonicity.* The total number of monotonic (increasing or decreasing) and nonmonotonic functions was determined and the percentage of monotonic curves was used as an index of the consistency with which subjects viewed the change in risk over time. Strict monotonicity was not required; that is, a monotonically increasing curve could be flat in places, but never go down.

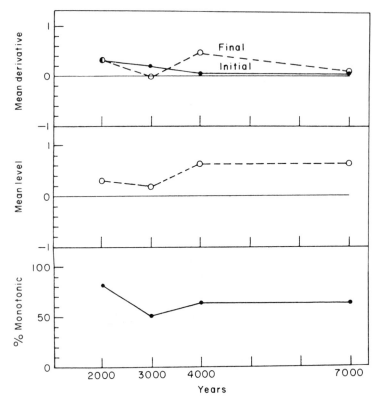

FIGURE 2. Risk profile associated with food as a function of the upper time limit (year). For more details see Figure 1 and the text.

(2) *Relative level.* This refers to the height of the curve at the longest time on the x axis for which there was a y value, as compared with the height of the curve at 1982. The relative height was coded as either $+1$, 0 or -1 depending on whether the final level was greater than, equal to, or less than the current value of the curve. The data then were averaged ($N = 16$) to obtain an index for the group that reflects a general view of the relative risk in the distant future compared to the present. This index may be seen as an 'optimism' indicator—numbers near $+1$ suggest low confidence (higher risk) in the ability of society to control the factors determining risk, numbers near -1 suggest a more optimistic attitude toward the future.

(3) *Initial derivative.* The subject's first impression of the future is reflected by the direction assumed by the curve in the time interval immediately after 1982. The initial slope (derivative) was judged as either positive ($+1$), flat (0) or negative (-1), and these data were averaged over all curves to obtain an index ranging from -1 to 1 (positive values associated with increased perceived risk).

(4) *Final derivative.* The direction of the curve during the last time interval was judged and coded as positive ($+1$), flat (0) or negative (-1), and averaged in the manner noted previously for the initial derivative. This index reflects a subject's 'last word' about the direction of the risk curve beyond the time scale available on the graph.

(5) *Termination point.* A 'doomsday' prediction appears in either of two ways: the

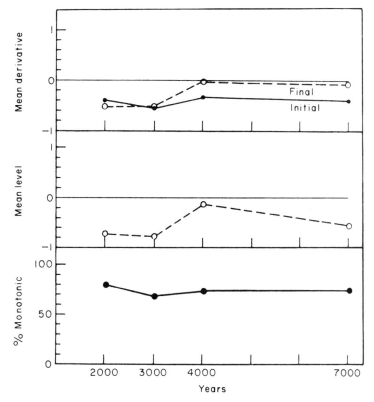

FIGURE 3. Risk profile associated with water as a function of the upper time limit (year). For more details see Figure 1 and the text.

height of the curve is everywhere greater than 0, but does not continue over the entire time scale; or the curve intersects the x axis (risk = 0). The percentage of 'doomsday' predictions was assessed for each item. Interviews with subjects after the experiment indicated that these cases represented genuine predictions of dire consequences, rather than a reflection of society's ability to solve the problem at that point in time.

Profiles of each item, based on mean values of the first four indices, are presented in Figures 1 to 9. In all cases the index is plotted as a function of the upper bound of the time scale (year). Hence, each point summarizes the 16 curves produced by a single group of subjects.

The first three figures show profiles for population, food, and water. A full description of the graphs in Figure 1 (population) will serve to illustrate the way these results should be interpreted. The lower (c) graph shows the percentage of monotonic curves as a function of the upper time bound. With the exception of the immediate future (2000) there is an approximately equal split between monotonic and nonmonotonic curves.

The middle graph (b) shows the mean level of the end point of the curve in respect to the starting point. The slightly positive values suggest that subjects perceive an increase in risk level at the most distant point in the future as compared to the present, though the effect is less pronounced for longer time scales. The top

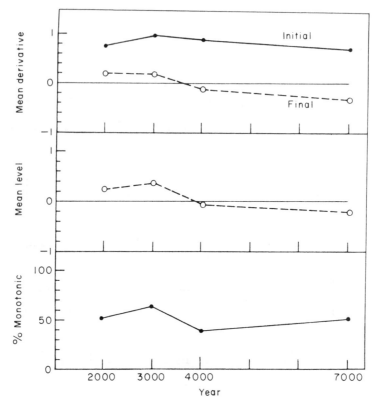

FIGURE 4. Risk profile associated with environmental pollution as a function of the upper time limit (year). For more details see Figure 1 and the text.

graph (a) presents both the initial and final derivatives (mean values). The initial increase in risk over some arbitrarily small time interval is usually greater than the final increase over a comparable interval, since the initial derivatives are closer to 1 than the final derivatives. In sum, subjects perceive a risk associated with population problems that becomes more serious in the immediate future but declines somewhat over longer time scales.

The profile for food (Figure 2) is essentially the same as for population. Approximately 50% of the curves are monotonic, the level of the curves is higher at the end of the time scale than at the beginning, and the derivatives tend to be positive. Unlike population, the perceived risk associated with food is closer to the zero line indicating no change over the upper time limit for both the initial and final derivatives.

The results for water (Figure 3) portray quite another situation. There are more monotonic curves, implying higher confidence in the predictions, the level of risk decreases over the course of the time scale for all upper time bounds, and the derivatives tend to be negative (especially the initial values). The perceived risk associated with water is clearly less than that associated with either food or population. In all three instances there are only minor fluctuations in the indices as a function of the upper time-boundary. This implies that subjects' predictions are the same for very long time-frames as they are for the immediate future.

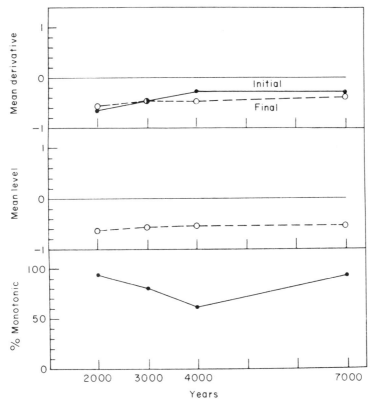

FIGURE 5. Risk profile associated with disease as a function of the upper time limit (year). For more details see Figure 1 and the text.

The succeeding four figures deal with health factors. Figure 4 gives the results for environmental pollution. The monotonicity index hovers around 50%, as was true for the previous hazards discussed. However, the mean level of the curves is quite nearly the same at the end of the time scale as it is at the beginning (b), and there are clear differences between the initial and final derivatives. Subjects' first reaction appears to be a pessimistic one (positive derivative) but matters stabilize (in respect to risk changes) by the end of the time scale, as indicated by the finding that the final derivative is close to 0. When coupled with the result that 50% of the curves are nonmonotonic, the separation of the initial and final derivatives suggests that a high proportion of subjects may have predicted an initial rise in risk followed by a fall and eventual leveling off at about the current level. In brief, the view expressed is that risk from environmental pollution will get worse in the near future but will improve relative to that peak in the long-term.

Quite a different picture emerges when subjects predict the risk associated with disease (Figure 5). The majority of curves are monotonic, the final level of risk is substantially less than the starting level, and both derivatives are negative for all time scales.

This optimistic view of our ability to reduce the threat of disease in both the immediate and long-term future is mirrored by the results for genetic defect (Figure

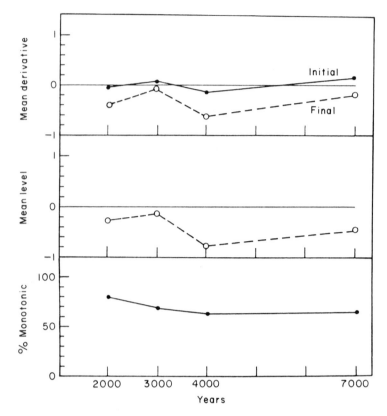

FIGURE 6. Risk profile associated with genetic defect as a function of the upper time limit (year). For more details see Figure 1 and the text.

6), although there are somewhat fewer monotonic curves and the initial derivative is very close to 0.

On the other hand, the same optimism is not apparent when assessing risk from mental disorder (Figure 7). The high percent of monotonic curves together with an unchanging mean level might indicate that subjects expect little improvement in the area of mental health over these long time scales. The negative values of the initial derivative imply, on the other hand, on optimistic attitude toward the short-term problem, a feeling that is not matched by the long-term prospect (the final derivative is everywhere positive). Although subjects predict a dramatic reduction in risk due to disease, it is evident they do not perceive this improvement to be equally likely for all subcategories.

Figures 8 and 9 give profiles for the perceived risk associated with nuclear power and nuclear war. In both instances the level of risk is predicted to remain fairly constant when the initial and final values are compared (b). However, the initial derivative is positive in both cases (increase in perceived risk) and the final derivative is lower and close to 0.

The percentage of monotonic curves is relatively constant at a level somewhat above 50% for nuclear power, but there is a large decline in this percentage for the two largest time bounds for nuclear war. Many of the curves for the latter hazard

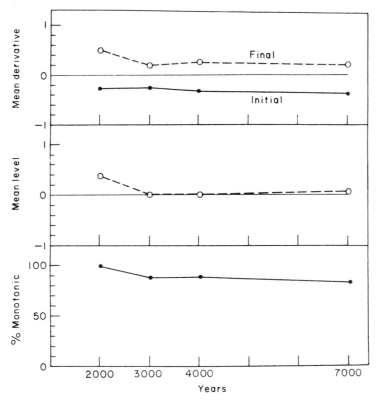

FIGURE 7. Risk profile associated with mental disorder as a function of the upper time limit (year). For more details see Figure 1 and the text.

were nonmonotonic in the sense that they rose over the early time intervals and then declined sharply, either by leveling out at about the same height as the current value (1982) or by plunging to 0 ('doomsday' prediction).

The percentage 'doomsday' predictions are given in Figure 10 for each of the items, with the upper time-boundary as a parameter. The greater the upper boundary the greater the number of 'doomsday' predictions; 38% of the subjects produced such predictions for population, food, and nuclear war for the year 7000 scale. These same items yielded percentages of 19, 12 and 12 for the shortest time boundary (2000). The lowest percentages were obtained for genetic and mental defects. Apparently, subjects made no attempt to achieve logical consistency in drawing curves for different items. For example, a 'doomsday' prediction concerning nuclear war would imply a similar fate for the remaining items, but this logic is not reflected in the data.

Discussion

The curve-drawing method appears easy to use, and though preliminary, a number of meaningful indices can be obtained to characterize entire sets of curves. Gardiner and Edwards (1975) have employed a similar method in their work on multi-attribute

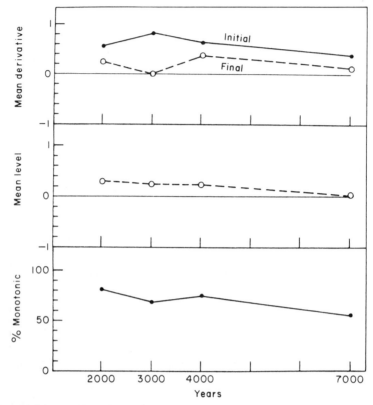

FIGURE 8. Risk profile associated with nuclear power as a function of the upper time limit (year). For more details see Figure 1 and the text.

utility measurement, but their purposes and analyses are quite different. The technique would seem to hold special promise for securing rapid estimates over continuous time scales. A simple and obvious improvement on the method would be to use a computer monitor that allowed subjects to draw curves directly on the screen. This would add precision to the data analysis as well. In addition, more care needs to be taken in future work with regard to the instructions. Perhaps some independent means should be used (e.g. a questionnaire) in order to determine the meaning subjects attach to such terms as 'predicted risk', 'perceived risk' and the like. Because of such methodological concerns, the present results, though provocative, must be considered tentative.

Turning now to the major results, the most striking feature is the invariance of risk predictions over changes in the upper time limit. An individual's view of the course of events over the next 20 years is almost the same as his or her view of events over the next 5,000 years. The result is consistent with a conjecture offered by Boniecki (1980) that '10–15 years seems to be the most distant practical horizon that the contemporary Western man may see as related to his own life experience' (p. 174). Not surprisingly, there does appear to be a slight increase in the number of nonmonotonic curves with the more distant time boundaries, coupled with an increase in the number of

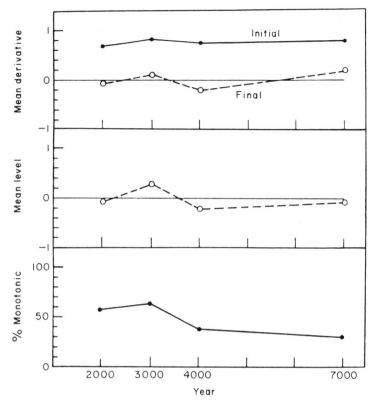

FIGURE 9. Risk profile associated with nuclear war as a function of the upper time limit (year). For more details see Figure 1 and the text.

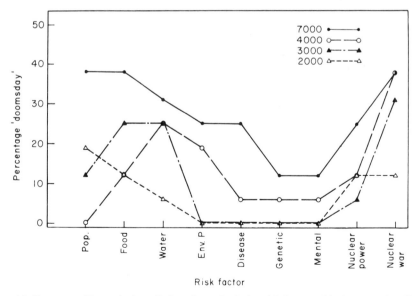

FIGURE 10. Percentage 'doomsday' predictions for each of nine risk factors with the upper time limit as a parameter. Each point is based on 16 curves. For more details see the text.

'doomsday' predictions. Approximately 40% of the subjects tested apparently think that human existence on this planet will end within the next 5,000 years, probably as a result of our involvement with nuclear power or war.

Quite probably the underlying factors determining an individual's view about these rather global issues is the degree of exposure to relevant news stories (Slovic *et al.*, 1982). With this possible explanation in mind it looks as though the medical profession has managed to impress this particular college population with humanity's ability to control suffering and disease in both the short and long term. The more pessimistic prognosis for population, food, environmental pollution and nuclear involvement can probably also be traced to the news media. Apparently, not as much attention has been given to potential hazards associated with the earth's water supply, since subjects were somewhat more optimistic about this issue. Naturally, it is not known whether the attitudes concerning risk are or will be paralleled by appropriate behavior directed to improving the global circumstances assumed to be responsible for actual environmental risks.

Acknowledgements

I appreciate the sound and cheerful advice of Marita Fairfield in conceptualizing this experiment. The article is based on a paper presented at the International Congress of Applied Psychology, held in Edinburgh, Scotland in July of 1982.

References

Boniecki, G. (1980). What are the limits to man's time and space perspectives? Toward a definition of a realistic planning horizon. *Technological Forecasting and Social Change*, **17**, 161–175.

Gardiner, P. C. and Edwards, W. (1975). Public values: multiattribute-utility measurement for social decision making. In M. F. Kaplan and S. Schwartz (eds), *Human Judgement and Decision Processes*. New York: Academic Press.

Kates, R. W. (1976). Experiencing the environment as hazard. In H. M. Proshansky, W. H. Ittelson and L. G. Rivlin (eds), *Environmental Psychology: People and their Physical Settings* (2nd Edit.). New York: Holt, Rinehart and Winston.

Kluckhohn, F. R. and Strodtbeck, F. L. (1961). *Variations in Value Orientations*. Evanston, Illinois: Row, Peterson.

Slovic, P., Fischoff, B. and Lichtenstein, S. (1982). Facts versus fears: understanding perceived risk. In D. Kahneman, P. Solvic and A. Tversky (eds), *Judgment Under Uncertainty: Heuristics and Biases*. New York: Cambridge University Press.

Svenson, O. Time perception and long term risks. Unpublished manuscript, Department of Psychology, University of Stockholm, 1985.

de Volder, M. (1979). Time orientation: A review. *Psychologie Belgigue*, **19**, 61–79.

SOURCES OF EVALUATION OF NUCLEAR AND RENEWABLE ENERGY CONTAINED IN THE LOCAL PRESS

RUSSELL SPEARS,*† JOOP VAN DER PLIGT‡ and J. RICHARD EISER†

† University of Exeter and ‡ Free University, Amsterdam

Abstract

This study investigated the sources of evaluative coverage concerning nuclear power and renewable alternatives contained in local daily press coverage. Ten categories of source were defined for their relevance to the nuclear debate and energy issues generally. Out of these, only 'pronuclear industries' and 'national government' produced more positive than negative appraisals of nuclear power. However, detractors of nuclear power were more varied, the most prolific category being the general public. Moreover, these trends were more marked for coverage in areas confronted with a possible new nuclear development. By contrast, alternative technology received far more positive appraisals and this was spread across a number of sources. Overall, whereas nuclear power was largely negatively evaluated and its support was mostly 'home-based' the reverse was true for coverage of alternatives.

Introduction

During the sixties and early seventies the continued expansion of nuclear power seemed assured. However, more recently, many factors have contributed to the turning tide of public opinion against this technology (cf. Kasperson et al., 1980; Nealy et al., 1983; van der Pligt, 1982; Thomas and Baillie, 1982 for reviews of public opinions trends). Despite this fact, the U.K. government and electricity supply industry seem as committed as ever to extending the nuclear option. It was against this background that, in an earlier paper, we compared local newspaper coverage of nuclear power and renewable alternatives (Spears et al., 1986). We found that nuclear power was overwhelmingly negatively evaluated on a number of different dimensions known to characterize energy issues (cf. Thomas et al., 1980), tending to reinforce popularly held fears about this technology. Given the widening gulf between policy makers on the one hand, and the public and the local press on the other, it seems important to distinguish between different 'sources' of evaluative coverage of nuclear power. (By 'sources' we mean those people, agents or institutions who are responsible for making or producing evaluative assertions about nuclear power, as reported in the local press.) The present study therefore seeks to extend the original content analysis by Spears et al. and differentiate evaluative press coverage according to its source. (In fact these two studies were conducted in conjunction with each other and on the same sample of local newspaper articles.) As in our earlier paper, the present content analysis retains information concerning the dimensions of evaluation variously used by different sources, and coverage of nuclear power is

* R. Spears is now at the University of Manchester, Department of Psychology, Manchester M13 9PL. This research was supported by an ESRC studentship to the first author and Exeter University Research Fund.

also compared to that for renewable alternatives. Finally we compare the distribution of sources of evaluation between areas which are confronted with the prospect of a new nuclear development, and those that are not. Spears *et al.* (1986) found that coverage of nuclear power was more evaluatively negative in the 'threatened' areas.[1] It is therefore important to determine whether the distribution of sources responsible for this pattern varies also. In summary, the present study discriminates the sources of evaluative coverage concerning nuclear power and its renewable alternatives contained in local press coverage.

Design: Method and Procedure

Sampling frame
The sampling frame was made up of all the local daily newspapers in the U.K. ($n = 103$). Any article referring to nuclear power and/or alternatives appearing in the first half of 1981, of more than one inch column space, was selected for coding. The overall sample was divided into two: the 'affected' and 'unaffected' subsamples. The affected sample corresponds to coverage in areas confronted with the prospect of a new nuclear development and comprised cuttings from four newspapers: *East Anglian Daily Times*, *Western Morning News*, *Evening Star* (Ipswich) and *Dorset Evening Echo*. The Suffolk papers qualify for this category because of the Central Electricity Generating Board plans to build a pressurised water reactor (PWR) at Sizewell. (At the time of sampling, the Public Inquiry had not yet started.) The other two are included because of the CEGB proposal, announced in 1980, to build a new nuclear power station in South West England, of which three sites in Cornwall and two sites in Dorset were put forward. Although this is a small number of papers compared with the unaffected sample, these four are among the five that are most prolific in terms of articles detected and comprise just over 30% of the sample as a whole ($n = 289$ for the affected sample; $n = 650$ for the unaffected sample).

Coding frame
Apart from recording the date and origin of articles, coding involved two distinct stages corresponding to questions addressed by this study and our earlier paper (Spears *et al.*, 1986). Specifically, whereas the previous paper was concerned solely with the evaluation of the technologies in press coverage, the present objective is to determine the 'sources' of these evaluations. 'Source' was operationalized as any 'identifiable *person*, *agent* or *institution* who, as far as could be determined from the text, was responsible for the particular evaluation in question'. In order to appreciate

[1] This concurs with attitude research which shows that support for the construction of a nuclear power station in one's locality tends to be much lower than support for the expansion of the industry in general (the so called not-in-my-back-yard effect; cf. Baillie and Thomas, 1982; Eiser *et al.*, 1986). Moreover, this is true here even of communities which already have a local nuclear power station–a factor which has typically been associated with more favourable attitudes to this technology. Thus, although we found some evidence that people living in the vicinity of an existing nuclear power station in the South West were slightly less anti-nuclear than others (van der Pligt, *et al.*, 1986*a*) they were still generally opposed to a new nuclear development in their neighbourhood (van der Pligt, *et al.*, 1986*a,b*). Attitudes are also opposed in the Sizewell area (Warren, 1981) which again has an established reactor. (Indeed press coverage in parts of the South West and Sizewell correspond to the 'threatened' sample of our present study).

more fully the precise method of ascribing sources, it is necessary briefly to describe the first ('evaluation') stage in more detail as the two stages are interlinked.

Stage 1: Evaluation. The unit of analysis was the sentence. Each article was scored for the number of positive and negative statements it made about nuclear power and alternative technology (neutral or purely descriptive statements were ignored). In addition, these evaluative statements were classified according to a number of categories or 'dimensions' known to characterize energy issues. Specifically, Thomas *et al.* (1980) performed a factor analysis of belief items concerned with energy technology and found five underlying dimensions: economic, environmental, technological, future political ('indirect risk') and physical/psychological ('direct life risk'). Spears *et al.* (1986) further subdivided the economic dimension into 'short' and 'long term' economic costs/benefits, and also added an 'unqualified' category to account for general evaluations which did not fit into the scheme. To summarize then, this resulted in a coding scheme of 2 (nuclear/alternatives) × 2 (positive/negative) × 7 (dimensions of evaluation) cells, making 28 in all (e.g. technology, positive, nuclear; technology, negative, nuclear; technology, positive, alternatives, etc.) Sentences were simply scanned for evaluative content and an entry made in the appropriate cell. If more than one dimension was relevant within a sentence, each was coded (see Spears *et al.* (1986) for more detailed operationalization rules for the coding, and for the intercoder reliabilities of the seven evaluative dimensions).

Stage 2: Attribution of source. For each entry of an evaluation, a further decision was made to determine its 'source' (see above); only sources of evaluations were coded. Ten categories of source were devised to classify those people, agents or institutions involved in the nuclear debate and/or energy policy in general. These are as follows:

(1) Pro-nuclear industry and organizations (e.g. CEGB, BNFL).
(2) UK central government.
(3) Advisory institutions and commissions.
(4) UK local government
(5) Independent institutions (e.g. Universities, non-governmental research institutes, 'experts', etc.).
(6) Media: active press and TV.
(7) Independent inquiries.
(8) Public (e.g. personal/public opinion, letters to the Editor, etc.).
(9) Anti-nuclear pressure groups and organizations.
(10) Pro-alternative pressure groups and organizations.

Greater elaboration of these categories can be found in Appendix 1.

A coding scheme similar to that for the evaluation stage was devised such that it was possible to code up to two sources for any one of the 28 cells. If there were more than two different sources for a particular evaluative cell then the two most dominant sources in that article were recorded (although in fact such multiple sources per dimension were very rare). If the same source recurred in the same cell within an article, it was not scored twice. Cases where the source was unstated, ambiguous or difficult to determine were simply not coded. For some examples of how sources were coded, see Appendix 2.

Reliability of Categories

Intercoder reliability based on a subsample of 110 articles achieved an acceptable level of reliability on appropriate indices. The combination of technology (nuclear/-alternative), evaluation (positive, negative) and dimension, resulted in 28 categories in all. Of these 15 yielded significant Cohen's k values (11 at $P < 0.001$). The absence of significant k values in the remaining categories (notably the future/political risk dimensions) was due to violation of assumptions associated with computing Cohen's k (cf. Fleiss *et al.*, 1969). Here, because the number of cells in the intercoder matrix is not small (11×11), it is legitimate to use a more simple measure of reliability such as percentage of agreement (Fleiss *et al.*, 1969). On this index, all remaining categories achieved a very respectable level of reliability (78·3–100% agreement).

Results

Description of the sample

Out of the 103 local daily newspapers in the U.K., 89 produced articles which contained evaluative coverage of nuclear power and/or alternative technology. In the six-month monitoring period a total of 939 articles were detected (cf. Spears *et al.*, 1986).

Distribution of sources

Data reporting the distribution of different sources of evaluation for nuclear power and alternatives are presented in Tables 1 to 4. These tables depict the sources of positive statements concerning nuclear power (Table 1), sources of negative evaluations of nuclear power (Table 2) and likewise for alternatives (Tables 3 and 4). Each table provides a breakdown of the distribution of sources on each dimension of evaluation (short-term economic, long-term economic, environmental, etc.) for both subsamples (affected, unaffected). Data in these tables denote the percentage of articles in each subsample which contain sources on any given dimension of evaluation. (N.B. While the percentage values appear to be generally small, these are expressed as a function of the *overall* subsample and not just as a proportion of purely 'nuclear' *or* 'alternative' articles.)

From Table 1 we find that the 'pro-nuclear industry and organizations' category is consistently the most prolific source of positive appraisals of nuclear power. In fact on all seven dimensions of evaluation, the recurrence of this source is greater in percentage terms than all the other sources put together. This is true of both the affected and unaffected subsamples although the 'pro-nuclear' category is *most* prevalent as a source of positive evaluations in the affected sample (where there are more articles concerned with nuclear power *per se*; cf. Spears *et al.*, 1986).

Comparing this distribution of sources with the pattern for *negative* evaluation of nuclear power provides a sharp contrast (Table 2). First of all it is apparent that sources of negative appraisals concerning nuclear power are much more widely distributed than for positive evaluations. That is, all categories of source are responsible for *some* negative statements about nuclear power, and compared with Table 1, most are relatively prolific. In particular, the 'public' category is a very substantial detractor, especially in the affected sample. For example, the public

TABLE 1

Sources of positive statements concerning nuclear power (percentage of articles per sample in which sources appear)

Source	Sample	Short-term economic	Long-term economic	Environmental	Future/political	Technological	Direct life risk	Unqualified
1. Pro-nuclear industry/organizations	Affected[a]	4·5	8·3	9·0	2·1	4·8	12·5	20·4
	Unaffected[b]	3·1	5·8	5·2	0·8	4·6	8·2	12·8
2. U.K. government	Affected	0·2	0·3	1·4			0·7	3·1
	Unaffected		0·8	0·6		0·2	1·1	1·2
3. Advisory institutions/ commissions	Affected		0·3	0·3	0·3		1·0	0·8
	Unaffected		0·3	0·3		0·3	2·0	1·0
4. U.K. local government	Affected		0·3	0·2	0·2			0·8
	Unaffected		0·3					
5. Independent institutions	Affected		0·7	1·0	0·3	1·5	1·0	2·8
	Unaffected		1·8	0·9		1·0	1·1	1·7
6. Media	Affected	0·7	1·4	1·0			1·0	2·4
	Unaffected		0·3	0·5		0·5	0·6	0·6
7. Independent inquiries	Affected							
	Unaffected							
8. Public	Affected		0·7	0·7	0·7	0·7	1·7	4·2
	Unaffected	0·2	0·5	0·6		0·2	0·6	0·8
9. Anti-nuclear groups/organizations	Affected			0·7			0·7	
	Unaffected							
10. Pro-alternative groups/organizations	Affected							0·2
	Unaffected							

[a] n = 289.
[b] n = 650.

TABLE 2

Sources of negative statements concerning nuclear power (percentage of articles per sample in which sources appear)

Source	Sample	Short-term economic	Long-term economic	Environmental	Future/ political	Technological	Direct life risk	Unqualified
1. Pro-nuclear industry/organizations	Affected	4·2	0·3	3·5	1·0	2·1	3·1	3·1
	Unaffected	8·5	0·8	4·6	0·6	4·0	5·8	4·3
2. U.K. government	Affected	0·3	0·3	0·3			0·3	0·3
	Unaffected	0·2	0·2	0·6	0·2	0·2	0·5	0·8
3. Advisory institutions/ commissions	Affected	2·4	1·7	1·4		0·7	1·7	3·1
	Unaffected	1·7	0·9	2·5	0·5	2·0	3·2	4·5
4. U.K. local government	Affected	0·7		3·5	1·0	0·3	2·1	5·9
	Unaffected	0·6	0·3	1·5	0·6	0·6	1·5	1·7
5. Independent institutions	Affected	1·7	1·7	4·8	0·3	1·0	4·5	6·2
	Unaffected	1·4	0·6	2·2	2·0	1·5	3·4	3·4
6. Media	Affected	3·8	0·7	4·2	0·7	1·7	4·5	4·8
	Unaffected	0·3	0·3	2·0	0·8	0·6	2·5	1·2
7. Independent inquiries	Affected	0·7					0·3	
	Unaffected	0·3					0·3	0·2
8. Public	Affected	8·0	4·2	13·8	9·3	4·8	13·1	30·8
	Unaffected	2·0	1·1	4·9	4·8	0·6	5·8	5·5
9. Anti-nuclear groups/organizations	Affected	3·5	3·8	7·3	4·2	1·7	6·6	14·2
	Unaffected	1·8	1·5	4·0	2·9	0·6	3·5	5·2
10. Pro-alternative groups/organizations	Affected	1·0	0·3	1·0		0·7	1·0	2·1
	Unaffected	0·3	0·2	0·3			0·2	0·6

made general or unqualified criticisms of nuclear power in over 30% of all articles in this sample. They also questioned its safety both in terms of human life *and* the environment in over 13% of cases. In general, sources tend to cluster on these three dimensions, where evaluative statements proliferate (cf. Spears *et al.*, 1986). Compared with Table 1, the 'pro-nuclear' and 'U.K. government' categories tend on the whole to be *less* frequent sources of negative appraisals than positive ones, whereas the reverse is true for all other categories of source.

Table 3, which presents the distribution of sources of positive statements concerning alternatives, also provides a strong contrast to Table 1. Unlike the positive appraisals of nuclear power, the sources responsible here seem to be broadly spread and not just located in the most intuitively partisan group (in this case the 'pro-alternatives groups and organizations'). Indeed, in the unaffected sample, the 'independent institutions' category seems to be the most prolific source. For example, over 17% of articles in this sample received acclaim from this quarter on the technological dimension, with 10·9% and 9·5% of such articles containing similar support on the long-term economic and unqualified dimensions of evaluation respectively. In the affected sample however, the pro-alternatives category is consistently the most common source of positive appraisals. Again, the long-term economic, technological and unqualified dimensions seem to be the main areas of acclaim.

Sources of negative statements concerning alternatives (Table 4) are relatively few and far between, reflecting the general paucity of negative evaluations overall (cf. Spears *et al.*, 1986). As for sources of positive appraisals, the main categories responsible here are the 'independent institutions' and 'pro-alternatives' sources.

Summary and Discussion

The distribution of sources produces a fairly high degree of consistency across the seven categories of evaluation. As might be expected 'pro-nuclear industries and organizations' are by far the most prolific source of positive evaluations concerning nuclear power, irrespective of dimension. Conversely the general public exceed even the anti-nuclear movement as the most frequent detractors of nuclear power. That the 'public' contribution to the negative sources is greatest in the affected sample is consistent with the fact that public opinion is relatively more anti-nuclear in areas confronted with a possible new nuclear development, though not necessarily in those with an existing nuclear facility (cf. van der Pligt *et al.*, 1986a; Thomas and Baillie, 1982; Warren, 1981).

It is interesting, although perhaps not surprising, that the government's evaluations of nuclear power are more congruent with those of the pro-nuclear category (mostly positive) than the advisory institutions and commissions which it authorizes (mostly negative). Out of the ten categories 'U.K. government' is the only other consistent advocate of nuclear power besides the industry itself. On the other hand, opposition is much more broadly spread across a number of different groupings. By comparison, the distribution of sources for alternatives is very different to that for nuclear power. Support for alternatives is much more widespread, proliferating in the 'independent institutions' and 'pro-alternatives' categories in particular. Sources of negative evaluations are relatively few and far between.

Regarding the dimensions of evaluation, sources tend to express their criticisms

TABLE 3

Sources of positive statements concerning alternatives (percentage of articles per sample in which sources appear)

Source	Sample	Short-term economic	Long-term economic	Environmental	Future/political	Technological	Direct life risk	Unqualified
1. Pro-nuclear industry/organizations	Affected		0·7			0·3		1·4
	Unaffected	0·2	0·3	0·2		0·5		0·8
2. U.K. government	Affected					0·3		1·0
	Unaffected		2·2	0·2		1·2		0·8
3. Advisory institutions/commissions	Affected	0·3	1·7			1·0	0·2	0·3
	Unaffected	0·3	2·2	0·5		1·4		1·8
4. U.K. local government	Affected		0·7			0·3		0·3
	Unaffected	0·3	3·1			2·3	0·2	1·5
5. Independent institutions	Affected	0·3	3·5	0·8	0·2	5·2	0·3	4·2
	Unaffected	0·6	10·9	0·3		17·4	0·3	9·5
6. Media	Affected	0·8	3·1			3·8	0·5	2·4
	Unaffected		7·1	1·4	1·1	8·6		2·8
7. Independent inquiries	Affected							
	Unaffected			0·8	0·2		0·2	
8. Public	Affected	1·0	2·1	0·8		2·4	0·3	3·8
	Unaffected	0·5	3·8	0·3	0·3	5·7	0·3	4·3
9. Anti-nuclear groups/organizations	Affected	0·3	1·4	0·3	0·2	0·7	0·2	2·1
	Unaffected	0·2	0·5	0·3	1·0	0·5		0·6
10. Pro-alternative groups/organizations	Affected	1·0	7·3	1·4		9·3	0·7	4·8
	Unaffected	1·2	7·7	0·8	0·2	9·5	0·2	3·8

TABLE 4

Sources of negative statements concerning alternatives (percentage of articles per sample in which sources appear)

Source	Sample	Short-term economic	Long-term economic	Environmental	Future/political	Technological	Direct life risk	Unqualified
1. Pro-nuclear industry/organizations	Affected	0·7	1·0			1·4		1·7
	Unaffected	0·5	0·5					0·3
2. U.K. government	Affected			0·2				
	Unaffected	0·9	0·3					0·3
3. Advisory institutions/commissions	Affected	0·3						
	Unaffected	1·4	0·5	0·6		0·2	0·3	0·2
4. U.K. local government	Affected			0·3				
	Unaffected	2·0		0·6	0·2			0·9
5. Independent institutions	Affected	1·7	0·3			0·3	0·2	1·4
	Unaffected	2·8	0·7	0·2		0·7		3·4
6. Media	Affected	1·0	2·3			2·9	0·2	0·7
	Unaffected	1·8	0·3			0·3		0·9
7. Independent inquiries	Affected		0·5	0·6		0·5		
	Unaffected							
8. Public	Affected	1·7		0·3		0·3		1·0
	Unaffected	1·2	0·2	0·2		0·6		0·9
9. Anti-nuclear groups/organizations	Affected	0·3						
	Unaffected							
10. Pro-alternative groups/organizations	Affected	2·4	0·3	0·3		0·3		0·7
	Unaffected	3·7	0·3			0·6	0·2	0·9

of nuclear power in terms of danger to human life and the environment. The relatively positive representation of alternatives is focussed on the technological issues
and long-term economic benefits. A substantial proportion of evaluation is also
unspecified for both technologies (see Spears *et al.* (1986) for greater detail
concerning the dimensions of evaluation).

Overall, the present study provides further descriptive evidence for the polarised
nature of the nuclear debate. Whereas the government and electricity industry may
have had a positive attitude to nuclear power at the time the sample was taken, our
data suggest this view was not widely shared. The public and (not surprisingly)
various anti-nuclear groups seem particularly vocal in their criticism of nuclear
power, largely outweighing the pro-lobby. This finding contrasts with an American
study which showed that policy makers and corporate interests were more able to
control the media presentation of a controversial technology at the expense of
conservationists and locally affected inhabitants (Molotch and Lester, 1975).
These researchers suggest that the power and influence of such groups gave them
'differential access' to 'event making' in the national press. Clearly, it is difficult
to draw comparisons between very different contexts and technologies (Molotch
and Lester's study examined coverage of an oil spill). However, the fact that our
study examined the *local* rather than the national press could help to explain the
greater voice of public protest and anti-nuclear pressure groups. Because of their
local activity, such groups and 'opinion leaders' may themselves have differential
access to the local press and be more representative of the target audience addressed
by local newspapers than is the largely centralized pro-nuclear lobby. Such a
interpretation gains some support from our comparison of affected and unaffected
samples. However, a more detailed analysis, extending beyond mere content analysis
of media output, would be necessary to fully explicate the complex relation between
the different interest groups on the one hand, and the national and local media on
the other.

References

Eiser, J.R., van der Pligt, J. and Spears, R. (1986). Local opposition to the construction of a
    nuclear power station: Differential salience of impacts. *Journal of Applied Social Psy-
    chology* (in press).
Fleiss, J. L., Cohen, J. and Everitt, B. S. (1969). Large sample standard errors of kappa and
    weighted kappa. *Psychological Bulletin*, **72**, 323–327.
Kasperson, R. E., Berk, G., Pijawka, D., Sharaf, A. B. and Wood, J. (1980). Public opposition
    to nuclear energy: Retrospect and prospect. *Science, Technology and Human Values*, **5**,
    11–23.
Molotch, H. and Lester, M. (1975). Accidental news: The great oil spill as local occurrence
    and national event. *American Journal of Sociology*, **81**, 235–260.
Nealy, S. M., Melber, B. D. and Rankin, W. L. (1983). *Public Opinion and Nuclear Energy*.
    Lexington: D.C. Heath.
Spears, R., van der Pligt, J. and Eiser, J. R. (1986). Evaluation of nuclear power and renewable
    alternatives as portrayed in the UK local press coverage. *Environment and Planning A*,
    18, 1629–1647.
Thomas, K. and Baillie, A. (1982). Public attitudes to the risks, costs and benefits of nuclear
    power. Paper prepared for the joint SERC/SSRC seminar on research into nuclear power
    development policies in Britain.
Thomas, K., Maurer, D., Fisbein, M., Otway, H. J., Hinkle, R. and Simpson, D. (1980). A
    comparative study of public beliefs about five energy systems, RR-80-15. International
    Institute for Applied Systems Analysis, Laxenburg, Austria.

van der Pligt, J. (1982). *Nuclear Policy and the Polarisation of Opinion*. Paper presented at the Conference: Issues in the Sizewell B Inquiry. London, October.

van der Pligt, J., Eiser, J. R. and Spears, R. (1986*a*). Attitudes toward nuclear energy: familiarity and salience. *Environment and Behaviour*, **18**, 75–93.

van der Pligt, J. Eiser, J. R. and Spears, R. (1986*b*). Construction of a nuclear power station in one's locality: Attitudes and salience. *Basic and Applied Social Psychology*, **7**, 1–15.

Warren, D. S. (1981). Local attitudes to the proposed Sizewell B nuclear reactor. Report RE19, Food and Energy Research Centre, October.

Manuscript received: 14 *October* 1985
Revised manuscript received: 26 *September* 1986

Appendix 1

Categories of source with some examples

(1) *Pro-nuclear industries and organization* BNF Ltd.; CEGB (and regionals); Uranium Institute; UKAEA (& Atom); British Nuclear Forum; Nuclear Power Information Group.

(2) *U.K. Government* Prime Minister; Parliament; Secretary of State for the Environment; Department of Environment/Energy; Ministry of Agriculture; Government MPS/MEPs; Acts.

(3) *Advisory Institutions* Ad hoc Royal Commissions; Standing Royal Commission; Parliamentary Select Committees/Reports; Private Members Bills; National Radiological Protection Board; Health and Safety Inspectorate (nuclear industries), etc.

(4) *Local Government* County Council; Borough Council; City Council; Metropolitan Council; Planning Authorities; MPs lobbied by local government.

(5) *Independent Institutions* Universities; Research Institutes without tied commercial/ nuclear funding; Nuclear Information Network; International Commission on Radiological Protection; 'experts'; unions; other countries; schools, etc.

(6) *Media* Press/TV reporting in an evaluative capacity—e.g. editorials/leaders; Press Conferences; Exhibitions/Events sponsored by the press.

(7) *Independent Inquiries* Public Local Inquiry (PLI); Ombudsman; Planning Inquiry Commission; Court of Law; The Assessor; The Inspector; WATT Committee (or on Government funded).

(8) *Public* Section 29 parties and other interested persons; Individual objectors; Letters to the Editor; Ad hoc local pressure groups (e.g. Stop Sizewell B); Public/ people's opinion (opinion in general); MPs lobbied by constituents.

(9) *Anti-nuclear Pressure Groups* SERA; Network for Nuclear Concern; National Peace Council; NCCL; CND; Greenpeace;* Society for the Protection of Rural England;* Friends of the Earth;* etc.

(10) *Pro-Alternatives Groups and Organizations* Any alternatives industry (e.g. ETSO, branch of UKAEA); National Centre for Alternative Technology.

---

* These sources may alternatively be ascribed to Category 10 depending on the context of the evaluation (i.e. anti-nuclear vs pro-alternative).

Appendix 2

Example extracts from articles

   (1) Taken from *Southern Evening Echo*, Southampton, 23 May 1981

		Evaluation	*Source*
S1	Safe energy campaigners left a petition at 10 Downing Street calling for a phasing out of nuclear power stations and a decision against building any more.	Unqualified, negative, nuclear	10. Anti-nuclear pressure groups
S2	Many people are less firm in their views, but share the concern which centres on safety *both* long and short term.	Environmental, negative, nuclear/ Direct life risk, negative, nuclear	8. Public 8. Public
S3	The problem of what to do with old stations is most relevant.	Unqualified, negative, nuclear	No source
S4	Experts are tending to disagree more and more as to the number of stages to be followed in the deactivation of stations.	Technology, negative, nuclear	5. Independent institutions

   (2) Taken from *Burton Daily Mail*, Burton, 3 January 1981

		Evaluation	*Source*
S1	Norwegian scientists believe they have discovered a new technique for turning the energy of waves into electricity.	Technology, positive, alternatives	5. Independent institutions
S2	Several years of tests in a lake north of Oslo, which angered fishermen but intrigued scientists abroad, confirmed the institute's 'hopes that the system will work and be economically competitive, said Dr Mehlum.	Environmental, negative, alternatives Technology, positive, alternatives Long-term economic, positive, alternatives	8. Public 5. Independent institutions 5. Independent institutions

(3) Taken from *The Birmingham Post*, 19 February 1981

	Evaluation	*Source*
S1 The CEGB yesterday hit back at criticism of the Government's nuclear programme, saying a Commons Committee had failed to understand the basis of the strategy.	Unqualified, positive, nuclear	1. Pro-nuclear industry
S2 However, the conservation group, Friends of the Earth, said its campaign against expansion of nuclear power in Britain was vindicated by the report.	Unqualified, negative, nuclear	9. Anti-nuclear pressure groups

(4) Taken from *Shropshire Star*, Shrewsbury, 27 January 1981

	Evaluation	*Source*
S1 Television scientist Dr Magnus Pyke has lent his weighty opinion to a campaign to extol the virtues of solar heating.	Unqualified, positive, alternatives	5. Independent institutions
S2 Background to the campaign is a Press forecast that within 20 years half of the homes in Britain will be using solar energy.	Technology, positive, alternatives	6. Press

(5) Taken from *North Western Evening Mail*, Barrow-in-Furness, 28 May 1981

	Evaluation	*Source*
S1 Two legal actions in which damages are being claimed on behalf of former Windscale workers are to be heard at Carlisle Crown Court on 9 June.	Unqualified, negative, nuclear	8. Public
S2 In both cases it will be alleged that radiation to which they were exposed in the course of their work affected their health, and in one case led to a fatal cancer.	Direct life risk, negative, nuclear	8. Public

CHILDREN'S PERCEPTIONS OF NUCLEAR POWER STATIONS AS REVEALED THROUGH THEIR DRAWINGS

J. M. BROWN*, J. HENDERSON† and M. P. ARMSTRONG‡

Department of Psychology, University of Surrey, Guildford, GU2 5XH;
† Manpower Services Commission, Moorfoot, Sheffield, S1 4PQ;
‡ *Southborough Boys School, Hook Road, Surbiton. Surrey, U.K.*

Abstract

Children's own drawings were used to reveal their changing perceptions of nuclear power stations over time. In all, 213 different children from three schools in the South East of England completed drawings in May 1983, December 1985, and in May and October 1986. A content analysis showed that the 1986 post-Chernobyl drawings were more likely to contain chimneys, smoke, pipes, and cooling towers. They were less likely to show bombs or rockets, features of earlier drawings. The collective pattern of features in the drawings was elaborated through a multidimensional scaling technique, the Guttman–Lingoes Smallest Space Analysis (SSA–I). A rationale is given for the use of this methodology whilst reference is made to a number of theoretical ideas for an understanding of the social processes involved.

Introduction

This paper describes a longitudinal study of children's perceptions of nuclear power plant as revealed through their drawings, collected from May 1983 to October 1986. The timescale of the study includes the Chernobyl nuclear disaster and spans a period of considerable controversy in the U.K. following the occurrence of a number of leaks from the British Nuclear Fuels reprocessing plant at Sellafield and attempts to site radioactive waste. Within behavioural geography and environmental psychology there is a long history of studying children's environmental knowledge. This is often elicited by drawings (Hart, 1979; Downs, 1981; Matthews, 1985; Moore, 1986). In these instances, the environmental phenomena being studied were part of the children's direct experience, their homes, schools or favoured play areas. Our study required children to explain a complex phenomenon of which they might or might not have heard and are unlikely to have had any direct experience. The reasons for studying children's awareness of nuclear energy are twofold. First, children are emerging citizens whose futures are intimately bound up with the progress of nuclear technology. Second, they represent the cultural litmus paper by which contemporary images and manifestations of nuclear technology may be revealed.

It is beyond the scope of the present paper to give a detailed content analysis of the media coverage of nuclear energy in the years covered by our study. However, it is worth noting that, from 1985 onwards, television news items, documentaries and films mentioning nuclear weapons and civil nuclear power greatly increased, especially after the problems experienced by the Sellafield nuclear reprocessing plant in Cumbria and the Chernobyl disaster. In various studies we have conducted of adults' reactions to nuclear power, summarized in Brown and White (1987), we consistently found television and newspapers to be the most likely source of information about nuclear energy. Less than 25% of the public participating in our surveys indicated they had actually visited a nuclear power station.

In a study commissioned by the Birmingham Trade Union Resource Centre, Sinclair (1987) found that of 312 14–16-year-olds answering her questionnaires, 68% had seen a nuclear power station. Of these, 90% had seen a picture of one on television and only 8% had actually visited one in real life. As with the adults in our own surveys, most youngsters (76%) were most likely to find out about nuclear energy from television, fewer (27%) from newspapers, and fewest (18%) from school. The type of television programmes giving them information ranged from the news (35%) and documentaries (35%), to series such as Captain Scarlet, The A-Team, or films including Threads, The Day After, The War Game, Superman, James Bond or Mad Max (from which 29% claimed they gleaned their knowledge). Viewing figures (Gunter, personal communication) would suggest that 18% of the 10–15-year-old population saw The Day After and 8% Threads.

The scope and relative coverage of topics appearing in newspapers can be approximated by making a count of entries appearing in the British Humanities Index. This is an indexing service that itemizes the content of articles appearing in leading British newspapers and journals. Under the heading 'nuclear power' the kinds of subheadings and numbers of indexed articles are given for the period covered by the study. Entry terms can also be categorized as relating to 'military' and 'civil' nuclear power. Summary figures indicate that until 1986 military references dominated. During 1986, Britan's television and newspapers extensively reported a series of

TABLE 1
Number of nuclear power articles itemized by British Humanities Index 1981–1986

Entry term*	1981	1982	1983	1984	1985	1986†
Energy (c)	–	1	2	1	–	7
Energy industry (c)	37	26	20	26	15	12
Explosive (m)	2	1	2	4	15	–
Fallout shelters (m)	3	1	4	–	–	–
Power stations (c)	3	10	41	17	13	44
Reactors (c)	9	2	1	1	–	3
Warfare (m)	9	6	9	23	8	5
Waste (c)	2	4	10	6	9	11
Weapons (m)	85	66	90	34	15	20
Disarmament (m)	–	–	70	14	45	23
Accidents (c)	–	–	1	–	–	3
Fallout (m)	–	–	1	–	–	–
Tests (m)	–	–	3	–	21	14
Radiation (c)	–	–	–	–	–	14
Nuclear war survival (m)	–	–	–	–	–	1
Proliferation (m)	–	–	–	–	–	7
Total c	51	43	75	51	37	94
	(34%)	(37%)	(29%)	(40%)	(26%)	(57%)
Total m	99	74	179	75	104	70
	(66%)	(63%)	(71%)	(60%)	(74%)	(43%)
Overall total	150	117	254	126	141	164

* c, civil; m, military
† January to September only

leaks at British Nuclear Fuels' Sellafield nuclear reprocessing plant; an announcement to dispose of low-level radioactive waste in one of four English villages; a
highly critical parliamentary inquiry into the U.K.'s nuclear waste industry; and, in
April and early May, the Chernobyl tragedy. 1985 had been marked by programmes
commemorating the anniversary of the Hiroshima and Nagasaki bombs.

In a particularly innovative study, Ryder (1982) explored in some detail the
understanding that a small group of secondary-school children gleaned about the
Three Mile Island accident from one particular bulletin of the BBC's Nine O'clock
News. It was evident from his investigation that school was not the only place
pupils learnt about either science or politics (p. 307). He advances the 'cultural raft'
hypothesis. This states that children do not value a television programme or newspaper article or even a lesson in school for its elegance or authenticity, but for the
degree to which it can provide fragments to add to their 'rafts' of knowledge. These
fragments enable the raft to float or sink or even to be rebuilt.

This notion of differential assimilation of knowledge has been expressed in another
way by Osborne, Bell and Gilbert (1983) who provide a framework to explain how
children respond to science teaching. They found that children may

(a) reject the teacher's version of science and prefer their own intuitive view;
(b) misinterpret the teacher's science and inappropriately incorporate it with
their own views, or alternatively, simply fail to reconcile their own and the
teacher's science;
(c) accept the new view but keep it in isolation from present views (here children
develop a strategy of reproducing the new view for the teacher or the
examiner yet retain their intuitive viewpoint);
(d) accept, and logically and coherently incorporate, the teacher's science.

However, the point Ryder (1982) was trying to make is that nuclear energy is represented on television as both a technical and political issue, and both were reflected
in the explanations the young people provided of the news story.

This idea of diffusion of scientific knowledge has been elaborated by Moscovici
and Hewstone (1983) who propose that the consumption of news in scientific fields
variously takes place in the classroom, from television, films, or during coffee breaks
(p. 104). Moreover, during its diffusion, scientific knowledge not only becomes
detached from its parent method or discipline, but is transformed by the particular
communication channel. Individuals selectively respond to the information in order
to serve new, social purposes and goals. These authors suggest that when dealing
with the complex and unfamiliar, individuals attempt to reduce confusion or ambiguity by anchoring the unknown to its nearest approximation.

Ryder (1982) provides a graphic example of this approximation process (p. 254).
One young girl reading a TMI newsscript misread several words. Examples of re-
placement words are as follows: 'value' for 'valve', 'exploded' for 'exposed', 'reaction'
for 'reactor'. Ryder concluded that she seemed to read according to a kind of psy-
cholinguistic guessing-game principle, substituting the nearest word within her
command that has some degree plausibility.

The questions suggested by these theoretical ideas lead us to ask (1) how do
children represent nuclear power stations; (2) can their representations be related to
images prevalent in the media; and (3) what functions do the representations serve
for the children.

Gaining access to representations of nuclear power had proved difficult in our studies with adults. We tried a sorting task in one study (Henderson, Brown and Spencer, 1984). Respondents were asked to pick out, from a set of 29 cards with various descriptions written on them, as many (or as few) descriptions as they wished that they thought typified a nuclear power station. Whilst our respondents were able to accomplish this task, this method suffers the weakness of providing the stimulus material. With the children we wished to have a more open-ended procedure. There is strong evidence to support the view that children may be restricted by verbal explanations and that drawing provides a possible alternative (Matthews, 1985). Chambers (1983) developed a methodology using drawings in his study which attempted to elicit the cultural stereotype of the scientist. He concluded that drawing avoided linguistic barriers and enabled comparisons to be made between groups of different languages and abilities. His draw-a-scientist test proved easy to administer, although he points out there are difficulties in coding and interpreting the drawings.

Crook (1985) argues that 'it is widely recognised that the content of children's drawings may provide insight into their feelings and thoughts about the world' (p. 251). In particular, Ryder (1982) found that drawings provided additional insights into understanding how the youngsters in his study explained the processes involved in the Three Mile Island accident.

Method

Data collection

A total of 213 children completed drawings of a nuclear power station. Drawings were collected, in May 1983, December 1985 and in May and October 1986, from three different schools in the South East of England. School A was a single-sex boys' secondary, B its sister girls' school, whilst school C was a girl's grammar school. All are located in Surrey within similar catchment areas. Details of the sample population are given in Table 2.

TABLE 2
Details of sample

Time of data collection	School	Age group (*years*)	Number of girls	Number of boys	Sub-totals
Pre-Chernobyl					
May 1983	A and B	11–12	39	37	
		13–14	14	14	
December 1985	A and B	11–12	3	17	124
Post-Chernobyl					
May 1986	A	12–13		32	
		14–15		17	
October 1986	C	12–13	24		
		14–15	16		89
Sub-totals			96	117	213

A. Secondary boys' school; B. Secondary girls' school; C. Girl's grammar school

Procedure

The children completed their drawings during a regular art lesson. Their usual teacher instructed them to draw a nuclear power station. No details, examples or reference material was made available. The children had not been forewarned that this would be their lesson topic. They chose their own medium for their drawings, either pencil, charcoal, felt-tipped pens or paint. The teachers were asked to impress on the children that this was not a test, that there was no right or wrong answer, and they could draw whatever they liked. Although pupils were asked not to copy from their neighbour, it was almost impossible to prevent this. Where it was obvious that drawings were virtually duplicates, we eliminated one from the analysis. However, it was not possible to control for ideas or inspiration gained by one child looking at another's work.

Although the same schools participated in the study, different children provided drawings, as it was felt that there would be a learning effect if the same child completed the task on two different occasions. The classes were also involved in discussions after they had finished their drawings. The researchers had the opportunity to discuss the drawings in detail as the children were completing them in one of the data collection sessions.

Data analysis

The drawings were content-analysed. Every feature of the individual compositions was recorded: for example, the presence of barbed wire, guards, helicopters, or particular words written on the drawings. Also recorded were the shape of the building(s) such as a 'natural gas container', a 'factory-like building', a 'tower that looked like the one damaged in the 1957 Windscale fire'. This 'pencil and paper' procedure generated an exhaustive list of features that was considered non-problematic. In other words, no attempt was made to reduce the items into categories at this stage.

Each drawing generated a profile in terms of the presence or absence of this total list of features. A frequency count of occurrences provided the basis of comparison over the period of the data collection. However, this decomposition of features gave no feeling for the style or collection of features appearing within drawings. We attempted several ways to sort the pictures into categories questioning art teachers, an art therapist and one group of students. A primary difficulty was that the drawings would not readily fit into mutually exclusive categories, and we abandoned this approach. As it proved impracticable to go back to all of the children who had previously participated in the study to help identify the types of drawing, we relied on a statistical procedure that imposed the least constraint on the data. We chose a non-metric multidimensional scaling technique, the Guttman–Lingoes smallest space analysis (SSA–I) (Lingoes, 1973).

SSA–I intercorrelates the column items and converts the measure of association into a linear distance such that the higher the association, the closer the points representing the items appear in the spatial plot. We used the Guttman–Lingoes association measure for dichotomous data, GLA1. There is a measure of goodness of fit between the correlations and linear distances called the coefficient of alienation. Guttman (1965) recommends a coefficient of 0·15 or smaller to indicate an acceptable spatial configuration of the correlation between items. Once we obtained a plot with an acceptable fit we used the correlations to help identify co-occurring features.

This procedure had been adopted by Muedeking and Bahr (1975) in their analysis of types of Skid Row behaviour; having no strong *a priori* grounds for defining patterns of behaviour, the authors used the SSA–I in a heuristic manner to determine distinctive types of Skid Row men.

Results

Most children drew building exteriors, chimneys or cooling towers with smoke, and dome-shaped or factory-like buildings. About a quarter of the drawings depicted guards, guns, barbed wire or other security devices such as closed-circuit television or arc lights. Words like 'danger' or 'keep out' also appeared. We compared pre- and post-Chernobyl data. Eight features were statistically significantly different (χ^2). Children were *more* likely to draw chimneys ($\chi^2 = 4.84$), smoke ($\chi^2 = 7.8$), a factory-

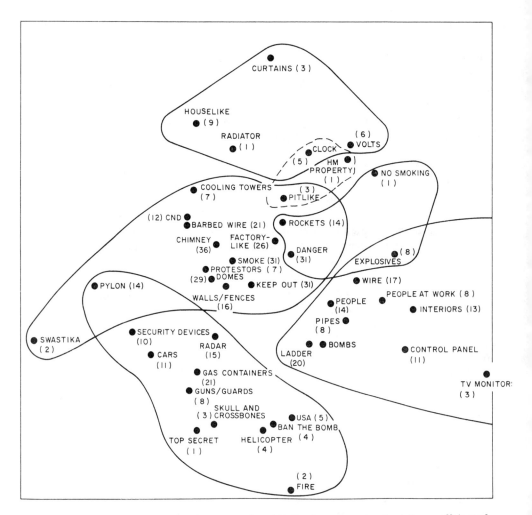

FIGURE 1. SSA of pre-Chernobyl drawings vectors 1 and 2 of a three-dimensional solution coefficient of alienation 0.15. Numbers in brackets represent raw frequencies.

like building ($\chi^2 = 15.73$), pipes ($\chi^2 = 15.63$) and cooling towers ($\chi^2 = 14.56$) after Chernobyl and the catalogue of disasters affecting the British nuclear industry. They were *less* likely to draw bombs or rockets ($\chi^2 = 10.77$), guards or gun emplacements ($\chi^2 = 7.79$) in their post-Chernobyl drawings.

The three-dimensional solution of the SSA–I of the 1983 drawings had a coefficient of alienation of 0.15. In the interests of parsimony only vectors 1 and 2 are presented. Also presented are vectors 1 and 2 of the four-dimensional SSA–I solution of the post-Chernobyl drawings, with its coefficient of alienation reaching 0.13.

The features of the children's drawings are positioned in the plots as a function of the likelihood of them appearing together in a drawing. Thus, items that appear closest together represent groupings of features found in the children's drawings. The lines drawn on the plots 'capture' features that tend to appear together in drawings. This is done with reference to the association coefficients calculated by

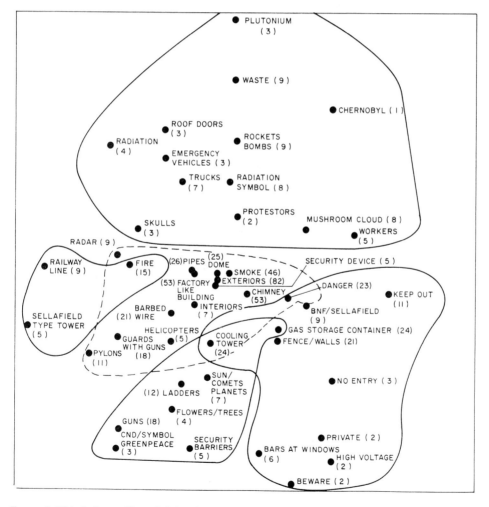

FIGURE 2. SSA I of post-Chernobyl drawings, vectors 1 and 2 of a four-dimensional solution coefficient of alienation 0.15. Numbers in brackets represent raw frequencies.

the SSA–I programme. Any pair of features having an association coefficient greater than 0·24 were included within the line of capture. It might seem that some items are close together but this is only an apparent spatial relationship (which can be verified by the association coefficient) because the particular plot represents only one perspective of a three- or four-dimensional configuration. Also, some elements may appear in more than one grouping indicating different types of drawings have some common elements.

The earlier drawings were of six basic types:

(i) domesticated settings with house-like buildings showing curtains at the windows;
(ii) industrial coal-mining settings with pit wheels;
(iii) interiors with control panels, explosives and bombs with people working at various tasks;
(iv) rockets and silos and danger signs;
(v) rockets amidst pylons, factory-like buildings and cooling towers;
(vi) suburban-type setting with gas-like containers, pylons and assorted graffiti, such as skull and cross-bones, 'ban the bomb', and radar dishes.

FIGURE 3. Power station drawn by pupil from School B in 1983

The 1986 drawings were marked by the loss of house-type drawings. There are five types:

(i) explicit references to the Chernobyl accident;
(ii) bombs and rockets amidst protestors and references to radiation;

FIGURE 4. Power station drawn by pupil from School C in 1986.

(iii) realistic-looking Sellafield-type establishment of British Nuclear Fuels;
(iv) juxtaposition of natural objects such as flowers and trees and guns or security barriers with a nuclear plant;
(v) high security buildings, cooling towers and gas storage containers.

Discussion

The children's drawings revealed a wide range of features and quite distinctive compositions. There was evidence of changes over time with different features appearing such as anti-nuclear missile protestors, radioactive waste, the Chernobyl disaster and the British Nuclear Fuels' Sellafield fuel reprocessing plant. This is a direct reflection of the contemporary media coverage of the issue. Interestingly, the Chernobyl disaster as such featured less in the 1986 drawings than might have been predicted given the extensive media coverage; rather, the drawings were more likely to contain chimneys, smoke, pipes and cooling towers. Rockets and bomb images were less prevalent in the 1986 drawings, and this can be contrasted with the drop in explicit military coverage of the issue in the media, overtaken as it was by the problems in the U.K.'s domestic nuclear industry and the Chernobyl accident.

The SSA-I provided a means not only to examine the constellation of images present in the drawings, but also the juxtaposition of features. Prior to May 1986, there is evidence of the children using prototypical environmental settings into which they transposed the power station—domestic, suburban, industrial and military. This is akin to the approximation principle of anchoring the unknown to the most plausible substitute. There was less coverage of the nuclear energy issue in the media, so the children had to draw from the surrounding cultural flotsam and jetsam to build their particular 'raft'. The later drawings indicated a shift towards, on the one hand, a greater realism, as in representations of BNFL's Sellafield works and, on the other, a sophisticated symbolic representation of the political dimension, with the presence of flowers and trees with CND symbols, security fences and danger signs. This is suggestive of the process described by Ryder (1982) whereby children acquire scientific knowledge. 'The pupil's enterprise is to build a raft, as he is carried along by the current, from every suitable piece of driftwood, oil drum or lashing that floats his way; and to replace them when they leak or rot . . .' (p. 16).

From discussion with the art teachers, it emerged that children not only resort to stereotypic forms in their drawings but also exhibit preferences for certain types of images irrespective of content. This has important implications for the content analysis of drawings since the objects and forms produced by the child may not have the significance attributed to them by others. Take, for example, the appearance of factory-like buildings with flat roofs. One teacher pointed out that the children's local area contained many factories of this type. This is possibly their only model for non-domestic buildings. It does not necessarily imply the children think nuclear power stations really look like this. The teacher speculated that, if children were asked to draw a coal-fired power station, they might have drawn the same thing. They had likened the unknown (or unobservable) nuclear power station to the nearest known (or observable) thing, a factory.

The lack of people in most of the 'power station' drawings suggested that a nuclear power station environment was seen as empty and alienating by the children. However, this may be an inaccurate interpretation. The art teachers we spoke to indicated that children find drawing people a struggle and tend to avoid including them unless explicity asked to do so.

Also present in the drawings were slogans such as 'top secret', 'danger', 'keep out'. Chambers (1983) found similar signs in his study of the image of the scientist, and concluded that the prevailing image was one of secrecy. Talking to the children revealed that secrecy is very much part of the representation of nuclear technology, as this extract from a boy in the pre-Chernobyl set illustrates.

> There is a bloke carrying a gun. The gun is to keep people out and in. If they get infected, you keep people in. They don't want everyone knowing what their secrets are, do they? They come in and out along these dirt tracks with nuclear missiles. Nuclear missiles get charged up in here—put the stopper on. They come out the gates and go off to Greenham Common or some other air base. The dome has some significance when they're charging up nuclear missiles—I think they've got boilers in there. There are first precautions and second precautions. First precautions are to keep everyone in there but if that fails, they'd probably be another squad of men line up there to get their guns out. That [pointing to the fence] would be made of wood but back up with mesh to stop people cutting into the base.

What is particularly interesting here is the presence of an armed guard to keep people both in and out of the power station, a meaning not possible to infer from inspection of the drawing alone.

Both inspection of the drawings and the use of statistical analyses reveal the physical content and spatial relationships of children's views of nuclear power plant. The symbolic meanings may be inferred to a limited degree but clearly the fullest understanding comes with the addition of a verbal explanation from the child. The drawings proved very effective in getting the children to articulate some notions they had about nuclear energy which we had not been able to elicit in pilot discussions using verbal (spoken or written) accounts only. However, the drawings are more than prompting devices, they have a potency of their own and convey their messages very powerfully. In some cases it is very obvious that the images are a direct copy of what the pupil has seen or read. In one instance a cartoon was reproduced from memory.

However it was also evident that some children had little or no idea of how to represent the power station physically or symbolically. For those who could, graphic skill was not necessarily related to actual knowledge or understanding. Thus some children performed the task in terms of what they imagined was required of them. Yet others produced conventional drawings that fulfilled their classroom obligations, but in talking to them it was clear they also maintained an elaborate and often fantastical explanation of the principles of nuclear fission as in the following example:

> [the power station] is on an island so no one knows where it is and its [got] solar panels so it gets more power into the building. The power is for the machines [because] they get the nuclear power somewhere else.

This is reminiscent of Osborne *et al.*'s (1983) findings that the children differentially assimilate science. Some children clearly had little idea of what a power station looked like and simply reproduced houses labelled 'nuclear power stations'. Others produced plausible, realistic representations, but their verbal explanations indicated they held to their intuitive view about the purpose of the power station or the process by which it produces power. Finally, we had examples of drawings and accounts that showed some understanding of electricity generation.

In conclusion, the drawings do reveal changing images of nuclear power stations over time which may be related to contemporary pictures and stories appearing in the media. Content analysis and statistical techniques help to elucidate the structure and dynamic nature of children's representations, but any explanation of the nuclear process must be augmented by the children's own accounts.

Acknowledgements

Our thanks are conveyed to all the children who participated in the study, to the teachers of the various schools, to Neil Ryder for his helpful observations and the comments of anonymous referees.

References

British Humanities Index (1981–82). London: Library Association.

Brown, J. and White, H. (1987). Public understanding of radiation and nuclear waste. *Journal of the Society for Radiological Protection*, **7**, 61–70.

Chambers, D. W. (1983). Stereotypic images of the scientist: the draw-a-scientist test. *Science Education*, **67** (2) 255–265.

Crook, C. (1985). Knowledge and appearance. In: N. H. Freeman and M. V. Cox (ed), *Visual Order: The Nature and Development of Pictorial Representation*, Cambridge: Cambridge University Press, 248–265.

Downs, R. (1981). Maps and mapping as metaphors for spatial representation. In L. Liben, A. Patterson, and N. Newcombe (eds), *Spatial Representation and Behaviour Across the Life-span*. New York: Academic Press.

Hart, R. (1979). *Children's Experience of Place*. New York: Irvington.

Henderson, J., Brown, J. and Spencer, J. (1984). The significance of environmental factors in appraisal of nuclear power. Paper presented to the International Association of People and their Pysical Surroundings. West Berlin, 27 July.

Lingoes, J. C. (1973). *The Guttman–Lingoes Nonmetric Program Series*. Ann Arbor: Mathesis.

Matthews, N. H. (1985). Young children's representation of the environment: a comparison of techniques. *Journal of Environmental Psychology*, **5**, 261–278.

Moore, R. (1986). *Childhood's Domain Play and Place in Child Development*. London: Croom Helm.

Moscovici, S. and Hewstone, M. (1983). Social representations and social explanations: from the 'naive' to the 'amateur' scientist. In: M. Hewstone (ed.), *Attribution Theory: Social and Functional Extensions*. Oxford: Blackwell. 98–125.

Muedeking, G. D. and Bahr, H. M. (1976). A smallest space analysis of Skid Row men's behaviour. *Pacific Sociological Review*, **19**, 275–290.

Osborne, R. J., Bell, B. F. and Gilbert, J. K. (1983). Science teaching and children's view of the world. *European Journal of Science Education*, **5**, 1–14.

Ryder, N. (1982). *Science Television and the Adolescent: A Case Study and a Theoretical Model*. London: Independent Broadcasting Authority.

Sinclair, P. (1987). Nuclear attitudes: an investigation into young people's attitudes towards nuclear power. Unpublished report to the Birmingham Trade Union Resource Centre.

HOW THE SIZEWELL B INQUIRY IS GRAPPLING WITH THE CONCEPT OF ACCEPTABLE RISK

TIMOTHY O'RIORDAN, RAY KEMP and MICHAEL PURDUE

University of East Anglia, U.K.

Abstract

There is no agreed level of safety that is 'acceptable'. Those responsible for managing and regulating safety believe that they can establish adequate parameters, but they recognize that those standards must meet with 'public approval'. This paper examines one way in which that approval is sought, namely the quasi-judicial examination of the merits of a proposal to construct Britain's first pressurized water reactor at the Sizewell B site on the Suffolk coast. The authors argue that the Sizewell B Inquiry appears to operate on the assumption that acceptable risk levels can be determined through argument and cross-examination. In its approach to the determination of acceptable risk the Inquiry seems dependent on professional judgement and expertise as assessed by legal minds trained to sift evidence. To ensure that all the necessary evidence comes before it, the Inquiry has established procedures to initiate the preparation and examination of expert viewpoints. The authors examine how far this approach is likely to command public confidence, and the extent to which the Inquiry pinned down the elusive concept of acceptable risk.

Aims of the Paper

The purpose of this paper is not to define acceptable risk, or even to discuss at any length how the concept of acceptable risk is approached. Its aim is to examine how one mechanism for gauging public approval, namely the public inquiry, grapples with the concept when endeavouring to assess the merits, on safety grounds, of a new nuclear reactor. As will be argued below, one of the purposes of the public inquiry, in Britain at least, is to command public confidence that a full and thorough examination of the technical information pertaining to a highly complex piece of engineering technology (in the particular case study illustrated here) has adequately been undertaken. In principle, then, the views of the Inspector to the Inquiry ought to reflect 'public approval'. We examine how far this is likely to be the case.

We also note that a clear distinction should be made between design targets for safety and the concept of acceptable risk. Engineers, statisticians and safety managers are continually devising better methods for predicting failures and faults in engineered systems and functioning machinery. They are also refining computer models of likely and possible accident scenarios. All this effort assists in the determination of safety targets which are built into the design of individual parts and systems of operation. These targets do not represent 'acceptable risk': they are critical design parameters vital to proving the safety case of a complicated piece of engineering technology. We do not address the issue of the concept and application of safety targets and probabilistic risk assessment except where these have implications for the definition and examination of acceptable risk.

One final introductory point: in this paper we look at risks which affect large numbers of people alive now or likely to exist, these are referred to as societal risks. A particular feature of the kind of societal risks discussed here is that the people exposed to these risks are not readily able to avoid them and, as individuals, they cannot be sure they can influence the process through which levels of public safety are reached.

Defining Acceptable Risk

We shall concentrate on three sources in this discussion, namely reports by The Council for Science and Society (1977, pp. 31–49), the Royal Society Study Group (1983, pp. 157–168) and the work of Decision Research (Fischhoff *et al.*, 1981; Fischhoff, 1983). There is a growing literature on this issue, but we feel these reports cover the relevant ground. All these sources argue that there are three categories of societal risk where societal risk is broadly defined as the joint probability of harm to a number of people who are unable to avoid the danger, and its likelihood of occurrence. There is a level where the level of danger and its likelihood is sufficiently manifest and serious as to give rise to general alarm: that risk is 'unacceptable'. Then there is a level at which the danger is either totally unknown, is ignored as being of no concern or is socially permitted (because it is familiar or regarded as unavoidable, or because the adjudged benefits accruing from the danger-inducing activity are calculated to be worth the possible harm). This category of risk is presumed to be accepted. Note that 'accepted' does not necessarily mean 'acceptable'. The Royal Society Study Group (1983, p. 157) make this point very clear:

> The fact that a risk is accepted is by no means a guarantee of its acceptability. Death has to be accepted by the individual, but is not acceptable to him. Moreover, acceptance may be associated with ignorance or a simple disbelief in the existence of any significant risk.

The third zone of risk, therefore, is this all-important area between 'unacceptable' and 'accepted' risk: it is in this zone that the concept of 'acceptable risk' is most applicable. To help in arriving at an understanding of acceptability we need to look again at the criteria normally applied to 'accepted risk'.

(a) That the danger is simply not known. This is an extreme case, but an important one for the nuclear industry. An accident or near-accident can dramatically alter perceptions of what is credible and what is not—both as to the characteristic of an accident or a general danger and to its likelihood. Accidents or the powerful imaginings of people not friendly to a particular risk source can remove the 'unknowability' aspect of certain kinds of risks.

(b) That the danger is so unlikely to happen as to give rise to no concern. This is a less extreme case, but is a central argument of the technical people who design nuclear plants and facilities. The problem here is that what may have seemed low enough to give rise to 'no concern' yesterday may not be so judged today. Public concern over both the safety record of the nuclear industry and public confidence in its assurances of competent management are changing to the point where the level of 'no concern' may be very elusive.

(c) That the risk is sufficiently low as to be 'worth it' in terms of the benefits that the risk-creating activity provides for society. This is the so-called risk-benefit calculation, currently being explored in attempts to incorporate cost–benefit analysis into risk management and regulation (see, for example, Webb, 1984; Nuclear Installations Inspectorate, 1984). This is a minefield, because there are explosive traps to the calculation of 'societal benefit' as much as there are to the calculation of 'societal risk'. One man's belief in the advantages of an additional kW of nuclear electricity may be counteracted by another's support for a reduced kW due to better utilization of electricity supply.

(d) That the risk cannot readily be reduced. People may be prepared to judge an intolerable risk as acceptable if they are confident that it cannot be reduced any further. Such a risk may be 'socially permitted' even though it is not tolerated. The problem here is that people no longer necessarily believe that a given danger is irreducible: advances in technology and scientific understanding together with recent developments in the art of regulation have opened up many new avenues for risk reductions.

We have left out one possible additional definition of acceptable risk, which could be regarded as an adjunct to (b) above. This is the risk comparability criterion, namely that what seems acceptable for some risk areas ought to be regarded as acceptable for others. Thus if society appears to be unconcerned about the probability of being struck by lightning (adjudged to be 10^{-6} per person per year) then society ought somehow to be unconcerned about the failure of a nuclear power plant resulting in an uncontrolled release of radionucleides should the designed probability be the same. The Royal Society Study Group (1983, p. 157) firmly rejected this approach:

> ... the existence, or indeed the informed acceptance of any set of pre-existing risks, provides absolutely no justification by itself for imposing any other hazard *even when the risk is reduced as far as it reasonably can be* [emphasis added].

The Royal Society tends to prefer the risk–benefit approach to its final judgement about acceptability (p. 158):

> ... The risk of an activity becomes acceptable only when it brings at least comparable benefits, and then only when the risk cannot reasonably be reduced further. This approach implies that there is not, and cannot be, a single acceptable level of risk.

The Council for Science and Society (1977, p. 36), however, takes this an important stage further. Acceptable risk should be that level of danger judged worthwhile *and deliberately chosen* by those exposed to it *in preference to feasible alternatives*. The Council adds an important new dimension here. Risks cannot be judged in isolation: they must be placed in a setting of looking at other means of achieving a given social objective, enabling some explicit form of societal choice.

The Council poses two ethical 'tests' for defining acceptability. The first is the right of the individual to protect personal interests. This is seen as too restrictive a position but, nevertheless, an important basis for looking at worker safety and compensation. The second is the 'public interest' alternative (Council for Science and Society, 1977, pp. 39–42) where acceptability is judged by some publicly credible and accountable process. The Council regards this 'test' as a matter for evolution in understanding and in communication between those responsible for controlling risks, those who are faced with the residual (i.e. regulated) risk and those expected to judge on the merits of the various (informed) views expressed: '... the scientific assessment of risks will evolve along with the other components of the dialogue, its methods and categories being improved by experience and criticism' (Council for Science and Society, 1977, p. 42).

Fischhoff and his colleagues would concur with this latter position, but would take the matter one stage further still. They conclude (Fischhoff *et al.*, 1981, p. 139) that the phrase 'acceptable risk' is not a unitary concept but a guide to how people make decisions amongst options. No risk–benefit calculation should be complete without reference to other approaches of achieving a mix of advantages and detriments in the same decision area (e.g. energy use). Even then, no approach to acceptable risk decisions adds much more than a fragment of all the beliefs and values relevant to forming judgements and determining priorities. Views about the 'acceptable risks' of nuclear power, therefore, cannot be isolated from analysis of preferences between a variety of electricity management options (including curtailing demand or using existing generating capacity more efficiently). It is misleading to separate discussion of acceptable risks into one risk arena alone, yet all too frequently this mistake is made.

We conclude this review of the definition of 'acceptable risk' by noting that the term has no absolute meaning or operational significance; that it cannot be applied to one risk arena in isolation; and is likely forever to change its meaning. Perhaps most important for our analysis is the view that 'acceptable risk' should only be regarded as a management and regulatory guide, and that its operational value can best be revealed through a mechanism for encouraging informed dialogue and public debate.

Examination of Acceptable Risk via Public Inquiry

We argue that to command public confidence, the mechanism for examining acceptable risk must be fair and thorough. By 'fair' we mean that all those with a genuine and relevant interest should be represented, and that where such interests are either not represented or under represented, then efforts should be made to ensure that their views are taken into account. By 'thorough' we mean that the concept of acceptability should be discussed in the widest possible context and that it should be the subject of detailed and expert examination—sufficiently detailed and expert to satisfy all interested parties.

The institution we examine in this paper is the statutory public inquiry. This device has a long history (for a review, see Kemp, 1983; Kemp *et al.*, 1984). Originally it was intended as a legislative mechanism to judge upon the merits of taking private land for a publicly orientated but private commercial interest—for example, land enclosure or the purchaser of rights of way for private railway, road and canal companies. As such it was a device of parliament aimed at providing busy legislators with a full view of the arguments involved. Subsequently the public inquiry has become less of a quasi-legislative institution and more of an administrative arrangement, advising ministers (or more latterly their senior officials) on the issues pertaining to a decision but leaving the wider political considerations aside. In essence, the modern public inquiry in Britain is a legally impregnated advisory mechanism for providing a recommendation, independently arrived at, for ministers to consider alongside various political matters of material significance which are not normally examinable by the inquiry proper.

The public inquiry has its parallels in other jurisdictions, but procedures differ markedly. For example in France, the *enquete publique* is a formal affair where objectors present their arguments to a tribunal chaired by a *Commissaire Enqueteur* (Macrory, 1982). They do not meet the proponent face to face, nor are they cross examined on their evidence. The *enquete publique* is one of a series of mechanisms for reaching decisions over major development projects in France so its limited scope for analysis and its excessive formality should not be seen as a fundamental failing in the larger pattern of reaching political choices. Nevertheless, the French experience to date does not provide the necessary conditions for the fair and thorough examination of acceptable risk.

In the U.S.A. and Canada the public hearing serves a similar function. However, this is an open forum, normally (especially in the U.S.A.) conducted through legal representation, where objectors and commentators can make their views known before a tribunal and can be cross-examined. Many public hearings have been criticized for their excessive dependence on lawyers and the expense of preparing voluminous documentation (see, for example, Nelkin and Pollack, 1982). As in France, the public hearing is not the only means by which objectors can challenge proposals: in the U.S.A. objecting parties frequently use the courts to attack the way in which decisions are made. Again we would submit that neither the hearing nor the courts are ideally suited for the fair and thorough examination of acceptable risk.

A recent variation to the public hearing in the U.S.A. is the round table discussion approach adopted by the Nuclear Regulatory Commission (NRC) to obtain public acceptance of its safety goals for nuclear power plants (Fischhoff, 1983). These goals are aimed at supplanting the myriad of specific regulations that the NRC has to make regarding the adequacy of engineered systems and safety procedures. The NRC held a series of workshops aimed at involving as diverse and as representative a cross-section of the public as possible. In a number of cases a sequence of hearings was convened in an effort to close some of the many gaps between competing views. Hopefully some form of consensus could thereby be achieved, resulting in a cadre of active support. The new NRC approach has yet to prove its worth, but it provides an interesting experimental example of the kind of approach advocated by the Council for Science and Society. Fischhoff (1983) is however wary of the achievements of the NRC approach, partly because the 'wider option' issue is not addressed, and partly because of the more technical implication that the NRC is in danger of making its goals too ambiguous and unenforceable in its efforts to command wider support.

We conclude this brief review of the public examination process by noting that procedural innovations are becoming increasingly evident. Bearing in mind that decision procedures vary enormously from country to country and that the relationship between political institutions and non-legislative procedures also varies considerably, it is significant to note that a common theme emerges of the need to design new institutional devices for determining acceptable risk, even though there are formidable problems involved.

There has only been one previous British public inquiry that specifically took evidence on acceptable risk. This was the inquiry into an application by British Gas to continue to use a methane terminal on Canvey Island, amidst a large and varied setting of chemical plants. In supporting the continuance of the plant, the Inspector (De Piro, 1983) drew as the comparison test, the possible additional risk imposed on the local population in comparison with other risks they face—from other industries, from coastal flooding and from general accidents. The Inspector based his judgement largely on the small additionality to societal risk imposed by the terminal, not on the characteristics of the risk that it imposed or upon alternative ways of handling methane gas in the area. As we shall see, the comparability criterion also played an important role in the Sizewell B Inquiry, but this was by no means the only criterion discussed.

The Sizewell B Inquiry

The Sizewell B Inquiry is being held under Section 2 of the Electric Lighting Act, 1909 into the application by the Central Electricity Generating Board (CEGB) to construct a pressurized water nuclear reactor (PWR) at the Sizewell B site on the east Suffolk coastline. The Sizewell A site is already occupied by a Magnox reactor completed in 1966. Under Section 2 of the 1909 Act, the consent of the Secretary of State for Energy is required, though in so doing the Energy Secretary can also grant *de facto* planning permission. The Inquiry is ostensibly concerned with the merits of this single application and the Inspector (Sir Frank Layfield QC) is only required to report to the Energy Secretary about that application. However, the CEGB have always made it clear that they regard the Sizewell B application as potentially one of a series (four to five) of PWR plant orderings.

In fact, the issues are much more complicated. The CEGB not only requires ministerial consent and deemed planning permission to build the generating station. It also must obtain a site licence from the Nuclear Installations Inspectorate (NII), the official nuclear safety and licensing authority located within the Health and Safety Executive. This licence will only be granted when the NII have completed their safety review. Ideally this process was supposed to have been finished prior to the onset of the Inquiry so that the CEGB and objectors could argue over an agreed design. But because of many delays in the preparation of the safety case for the proposed reactor, the licensing process is being run parallel to the Inquiry proceedings. In fact the Inquiry will be concluded before the licensing process is completed, so the Inspector has gone to great lengths to satisfy himself that the licensing process and administration of safety regulation are competently undertaken.

The Sizewell B Inquiry is ostensibly a non-political device for analysing what is in part a technological and in part a political issue. At stake is not only the character of future British energy policy and the likely shape of electricity generation for at least 25 years, but the credibility of the 'big' public inquiry itself. Can it really juggle the complexities of high science, high technology and high politics, and still be fair and thorough?

As we have hinted, this question raised the issue of legitimacy, namely the extent to which the Inquiry can reach or is seen to be able to reach a thoroughly independent conclusion given the wider political considerations involved. Legitimacy embraces both purpose and style: it extends to the degree to which the informed public are confident that the Inquiry is *authentic* in its independence from any form of political bias, and *faithful* to its reporting and appraisal of all the relevant arguments presented and examined before it, that is, the extent to which the Inquiry is both fair and thorough. These matters have already been discussed by Wynne (1982) and Kemp (1983) with regard to the Windscale Inquiry (into a proposal to establish a thermal uranium oxide reprocessing facility at Windscale, now Sellafield, in Cumbria). From different perspectives both authors concluded that the Windscale experience was neither fully authentic nor faithful in its function and style. Both felt that the legalism of that Inquiry prevented the full expression of 'unverifiable' but strongly felt feelings (notably about the safety of radioactive waste reprocessing and proliferation of nuclear weapons). Wynne also believed that the Inquiry was trapped by its particular style of analysis into reaching an inevitable conclusion to support the reprocessing plant. Kemp argued that the Windscale Inquiry was serving (possibly unintentionally) the particular interests of state nuclear policy rather than a wider, but more diffusely articulated, public interest.

The Sizewell B Inquiry is arguably on trial. It must persuade a potentially sceptical public that it is legitimate in terms of being fair and thorough to all parties involved in presenting argument and counter-argument, and to the issues before it, *irrespective of whether these issues have been raised by participating parties*. It appears that the Inspector and his four assessors (on economics, radiobiology, safety and irradiated fuel transport) are concerned that the Sizewell B Inquiry meets these requirements. The Inspector did write to the Energy Secretary to put the arguments of objectors that they should receive financial assistance to put their cases (due to the complexity of their evidence and the cost of obtaining adequate expertise). The Energy Secretary refused to grant financial aid: he argued that this would set an awkward precedent for other inquiries of this kind, and that the safety case was in any case being independently reviewed by the NII. This decision naturally upset the objectors some of whom refused to participate any further. Many of the rest indicated that the Inquiry could hardly be 'fair' if a consumer-financed public corporation with massive resources could only be opposed by groups dependent on charitable appeals for financial support.

The Inspector also requested and obtained ministerial permission to have his own Counsel, Mr Henry Brooke QC. Mr Brooke's role is to pursue inquiries on behalf of the Inspector and his Assessors not covered adequately by objectors or proponents, and to ensure that the complexities and mysteries of the specialized evidence were understood by informed lay people. Over the course of the Inquiry it has become evident that the Inspector, his Assessors and Counsel have taken an increasingly active role in following up matters that would not otherwise be fully explored, explained and analysed. This has been an evolutionary process, and more obvious in some aspects of the argument than in others. Nevertheless, the general trend is toward a more *investigatory* and *active* Inquiry, ready and capable of obtaining evidence on its own account and of cross-examining both proponents and objectors on matters of public concern (see Kemp *et al.* 1984; Purdue, *et al.* 1984).

Two examples of this investigatory element will be highlighted in this paper. The first applies to the seeking of guidance by the Inspector and/or his Counsel of a particularly knowledgeable witness as to how the Inquiry should approach a particular problem. This is an important development (though not especially unusual) because it suggests that the Inquiry is looking for assistance in developing criteria to assess the extensive evidence placed before it. It also suggests that the traditional adversarial style of examination may not be entirely appropriate for dealing with certain parts of this evidence, in order that the best answer can be obtained. The second example is the calling, by the Inspector, of special witnesses. These individuals are recognized experts asked to provide views and commentary from a neutral position, acting to assist the Inquiry over matters of technical complexity or specialized opinion.

In a sense, then, the Sizewell B Inquiry is trying to establish its own legitimacy by being more investigatory yet remaining faithful to its constitutional role of being advisory and recommendatory. This raises some fascinating issues with regard to its handling of 'acceptable risk'. First, the legalism of the Inquiry inhibits but does not eliminate the exploration of value-laden aspects of acceptability. This is because the legal approach encourages the Inquiry to seek judgements couched in experience, expert analysis and logical argument, not emotional reaction. Second, the paralleling of the official licensing process alongside scrutiny of both the safety case and regulatory procedures creates doubts both about 'fairness', since objecting parties have to respond to a constantly changing safety case, and 'thoroughness', since certain safety issues may remain unexplored in sufficient depth. Third, because of a tendency to politicize both the risk debate *and* the Inquiry, there is a possible tension between the use of the Inquiry to *explore* and to *examine* (its administrative, legal and investigatory roles), and the manipulation of the opportunities the Inquiry provides for particular political argument.

To a degree, therefore, the Inquiry is being caught in the crossfire which is now characteristic of the 'acceptable risk' debate. Its aim is to command respect through its independence and authority. Its tactics are to ensure the criteria of fairness, fullness and thoroughness are observed as completely as possible through the adoption of an investigatory style where the normal adversarial procedures do not appear to do sufficient justice to the issues being raised. The extent to which these developments prove successful remains to be seen: as we write (December 1984) the Inquiry has heard all the evidence, but the critical closing statements have still to be presented. So, in taking a look at how the Inquiry has handled the concept of 'acceptable risk', we can only make preliminary observations on its performance with regard to particular forms of rationality, the licensing procedures and procedural style. What follows is necessarily representative rather than comprehensive: the evidence, in the form of written proofs and extensive cross-examination, is so voluminous that only careful selectivity is possible. In addition, we avoid making final judgments on individual evidence and the Inquiry outcome. Commentary at this stage would be inappropriate.

Definition and Critique of Acceptable Risk

The evidence of the CEGB

The CEGB's position on acceptable risk was presented by Mr Roy Matthews (CEGB/P/1, 1982) Director of the Board's Health and Safety Division. He is responsible for determining safety criteria well below the thresholds set by the International Commission on Radiological Protection (ICRP) and its British counterpart, the National Radiological Protection Board (NRPB). These thresholds determine unacceptable levels of radiation exposure. The aim of the CEGB, as appraised by the NII, is to ensure that the risks to the public from normal plant operation are 'as low as reasonably achievable' (ALARA). This is a point worth noting. Unlike the American approach to nuclear safety, the British licensing authorities do not work to specific targets (outside of radiological standards as laid down) nor do they regard any given level of safety as immutable—even if designs subsequently have to be altered. These points are raised in the NII's evidence to the Sizewell B Inquiry (NII, 1983a) and elaborated in Addendum 2 to that Proof (NII, 1983b). These in turn are based in the NII's Safety Assessment Principles (NII, 1979). The customary U.K. regulatory practice may be described as 'monitored self-policing'. This is usually a matter of great pride amongst U.K. health and safety professionals: regulation is not imposed from without, it is supposed to be motivated from within the client organization.

It could therefore be argued in the first instance that the acceptability of nuclear risk is what the CEGB decides it ought to be. Mr Matthews (CEGB/P/2, 1982, para. 97) outlined for the Inquiry the well-established safety criteria followed by the CEGB. Note that these criteria are design target levels: as discussed earlier they do not necessarily equate with acceptable risk, though we detail below that in its evidence the Board effectively attempted to equate the two. It is also worth stressing that these targets apply to *uncontrolled releases of radiation*, and apply to *the most exposed individual*.

(a) Any single accident which could result in an uncontrolled release of radio-activity following the breach of some or all protective barriers should have a frequency of less than 10^{-7} per reactor year.

(b) The total frequency of all accidents leading to uncontrolled releases should be less than 10^{-6} per reactor year.

(c) For less severe accidents, the predicted frequency for whole body doses of 100mSv (10 rem) should not exceed 10^{-4} per reactor year. (This is in accord with guidelines set by the NRPB (1981), but the Board's interpretation of the ICRP/NRPB guidelines was challenged at length by the leading objecting organization on the Board's safety case, the Friends of the Earth (FoE) (Day 181, pp. 53–55).)

These criteria have the support of the NII, (1983b, p. 6) which comments:

> ... all accidents should be as remote as reasonably practicable. While the Inspectorate does not apply a rigid probability rule to them, *as a rule of thumb* any such fault sequence would be expected to be less frequent than about once in 10^7 year. [Emphasis added.]

The CEGB sought to justify its levels of design accident probabilities on the basis that there is a 'generally conceived acceptable risk of death of 10^{-6} per year' (CEGB, 1982, para. 14.44). Mr Matthews stated (CEGB/P/2, 1982, para. 98) that while 'there is no generally acceptable level of risk' the Board was following the views of 'various authorities'. The authorities cited include the International Commission on Radiological Protection (1977, paras 117–118), the Royal Commission on Environmental Pollution (1976, para. 171) the Advisory Committee on Major Hazards (1976, para. 19) and Lord Ashby (1978, p. 71). Each of these sources include statements to the effect that a risk of 10^{-6} per year 'is likely to be acceptable to any member of the individual public' (ICRP) or of 'no concern to the *average person*' (Ashby, 1978, p. 71).

There is some ambiguity in the Board's position over the link between the 10^{-6} 'acceptability' criterion. In its Statement of Case (CEGB, 1982, p. 72) the Board stated:

> The Design Safety Criteria provide guidance on important safety related factors which need to be taken into account during design. *Several of these are based on*

the concept of acceptable risk, and are expressed in numerical or probability terms
as design targets for each reactor on site, while others describe targets in qualitative
or engineering terms. [Emphasis added.]

But in answer to a question from Mr Brooke, Mr Matthews (Day 155, p. 77 C,D)
hinted that the 10^{-6} criterion was a professionally agreed target between the Board,
the nuclear industry and the NII. He qualified this remark by adding (Day 155,
p. 77 F) that

> ... these targets are a means of communication between my Department and the
> Board to the designer of what is needed. They are not intended as acceptance targets;
> they are not intended as acceptance targets for public risk, although in fact risk
> is obviously involved and there is some implication of the risk which we are expect-
> ing the public to accept.

The Board's position on 'acceptable risk' is based on two premises. One is that
the 10^{-6} per year yardstick will meet with *political* approval because specialist authorities argue that it is the correct target at which to base engineered safety features,
on the basis that the risk is 'negligible' in comparison with other dangers. The other
is that statistical techniques of probabilistic risk assessment (PRA) are so refined
that accident sequences which may (and indeed should) never happen can be calculated with sufficient confidence to command both scientific *and* public support.

Both of these are contentious suppositions. On the 10^{-6} issue, the nuclear industry
is by no means unanimous in its views. Chauncey Starr (1983, pp. 256–257) feels that
the industry should be seeking 'reliability and safety levels better than the "negligible"
level for the public (i.e. 10^{-6}) ... It is suggested that an acceptable upper bound of
potential nuclear risk is arbitrarily set at one-tenth of this level (10^{-7} per person per
year). This would put nuclear risks at the "negligible" level ...' Starr continued that
the U.S. Regulatory authorities should agree to this position and that the residents
near a nuclear situation should be given 'compensatory benefit sufficient to justify
accepting a risk equivalent to the normal risk of living. This is likely to be 1,000–10,000
times greater than the design target.'

The Royal Society Study Group (1983, pp. 179–183) was much more cautious.
The Group distinguished between different categories of risk, and specifically stated
that a common yardstick (for example, chance of death per person per year) would
be misleading: 'Risks and associated detriments should not be aggregated into some
single index unless the elements which contribute to that index can be disentangled'
(p. 182). The Study Group was also sceptical that any figure would be suitable
for the acceptability of a nuclear catastrophe described as a hazard 'very salient
in the public mind' where 'significant sections of the population may continue to
be concerned, whatever the level of risk, however small'. The Study Group concluded
that even at levels of apparent triviality to the designer (10^{-8} or even 10^{-9})
sections of the public and the media would still not be satisfied (p. 183).

The Evidence and Cross-examination by Friends of the Earth

On Days 162–164 the FoE counsel, Mr John Howell, closely cross-examined Mr
Matthews on the 10^{-6} yardstick. Part of Mr Howell's brief was a proof supplied
by Colin Green (1983). During Day 163, Mr Howell, drawing from FoE's interpretation of the Royal Society Report, elicited a number of responses from Mr Matthews.

Mr Matthew's Proof was completed some six months before the Study Group had published its findings, a point noted by Mr Matthews (Day 163, p. 10D), but the Board had every opportunity to furnish an Addendum, as it has done for most of its proofs since the Inquiry began. The following points emerged from exchanges on Days 163 and 181:

(i) that the Board had relied upon a single index of woe (i.e. death per person per year) when some multiple measure might have been provided (Day 163, pp. 6C, 16F, 48D) and that it had made no distinction between immediate and delayed deaths (Day 163, pp. 8A, 16A);

(ii) that the 10^{-6} index referred to revealed preferences which do not provide suitable guides for large consequence, low probability events of a highly salient character, where inter-risk comparisons are not appropriate (Day 163, pp. 13A, E. 17A, 20C);

(iii) that the reasoning behind the adoption of the 10^{-6} criterion by the four cited authorities (Day 163, p. 41C) had not been closely examined by the Board. Some of the sources were more than six years old (Day 163, p. 42E) and two of the sources (the Advisory Committee on Major Hazards and Lord Ashby) had stated that their thinking on this matter was tentative (Day 163, p. 71F–G; p. 77E–F);

(iv) that the Board relied on the negligibility criterion of acceptable risk (i.e. so low as to be of no concern) in interpreting the conclusions of each of the four cited authorities (Day 163, pp. 50H–51A); that the ICRP recommendations apparently favouring 10^{-5} to 10^{-6} referred only to routine, *not* uncontrolled releases (Day 181, pp. 53–55); and that according to the Royal Society Study Group a lower figure than 10^{-6} or 10^{-8} might be appropriate for a disaster such as a nuclear catastrophe (Day 163, p. 49G–H);

(v) that the fact that existing risks are acceptable does not in itself justify the imposition of more (Day 163, p. 53F) and that the Board did not take into consideration the possible benefits when establishing its 10^{-6} criterion (Day 163, p. 54A).

Whilst these exchanges must be read in their full context, they nevertheless reveal the power of well-briefed cross-examination (as wished for by the Council of Science and Society) for the thorough exposition of a party's case. (They further reveal the importance of keeping up to date with the state of the art on acceptable risk.)

On the witness stand, (Day 181) Mr Green reinforced these points. Under cross-examination by the CEGB's principal counsel, Lord Silsoe, Green was pressed to offer an acceptable risk target 'better or different from the Board's' (Day 181, p. 15B). Green was not briefed to do so but supported the view of another FoE witness, Kjellstrom (1983, p. 9) that 'ultimately safety criteria can only be established by political decisions', a view also shared by the Royal Society Study Group (1983, p. 143). Lord Silsoe also agreed to this but sought to show the Inspector that, in the absence of any other view, these political decisions would have to be influenced by the best scientific advice available, and, as he observed (Day 181, p. 28G), 'honestly come by'.

Green's only counter to this line of questioning was to argue that a range of reasons together with the relevant reasoning and underlying assumptions should be provided for the public to help them make more informed choices (Day 181, p. 30E). Lord Silsoe sought to suggest that this would not necessarily give any clearer guidance to politicians. To illustrate this point, he read aloud a letter from Lord Ashby to FoE (Day 181, p. 52E, F). In this note, Lord Ashby admitted that while it was a 'very questionable habit' to assume that comparisons can be made between different kinds of risks solely on grounds of their frequency, nevertheless the 10^{-6} target was still the 'rule of thumb for political decisions'. Lord Silsoe sought to reinforce the credo of the 10^{-6} target whilst attempting to manoeuvre Green's cogent proof of evidence into the sidelines.

The Evidence of Independent Witnesses

The Inquiry heard evidence from two witnesses, both recognized for their knowledge of risk assessment. The role of these two witnesses was, however, slightly different. Sir Edward Pochin, an emeritus member of the ICRP and an adviser to its British equivalent, the National Radiological Protection Board (NRPB), was asked by the NRPB to appear. The Inspector had invited the NRPB to produce a witness on the wider issues relating to public and occupational health and radiation. Technically speaking, then, Sir Edward was a witness for a neutral party to the Inquiry. The other witness whose evidence we examine was Professor Trevor Kletz, an authority on the application of risk analysis to the chemical industry. Professor Kletz was specifically invited by the Inspector to provide proof and be cross-examined. He acted not only in an independent but also in an advisory role.

We begin with Sir Edward Pochin's evidence. On Day 152 (p. 49G) Sir Edward was cross-examined by Mr Brooke on his views regarding the conclusions of Professor Eiser and his team at Exeter (see van der Pligt, 1985) namely that people's perception of the risk of nuclear energy was mainly determined by the characteristics of the hazard—notably its catastrophic and involuntary nature and the fact that it is an unknown hazard. According to this research the public were not biased or irrational, they simply defined risks in much broader terms than the expert.

In response Sir Edward (Day 152, p. 51E) admitted that the risks to the general public should be less than occupational risks (where workers allegedly voluntarily and knowingly accepted them), and emphasized the ICRP view that the 10^{-6} target was only 'likely to be acceptable' to any *individual* member of the public. He agreed with Mr Brooke and the Royal Society Study Group that comparisons of risk could be misleading because such comparisons suggested that what was 'acceptable' for one hazard was not necessarily acceptable for other hazards.

Yet the CEGB insisted that such comparisons were still useful. In his evidence Mr Matthews produced a table (CEGB, P/2, 1982, p. 78) showing different levels of risks in everyday life. This placed the 'radiation exposure risk' well down in the league table of danger. When cross-examined by Mr Brooke on this in the light of Sir Edward's remarks, Matthews answered (Day 155, p. 72E) that the table was published

> To give an order of indication of the risks that people run in everyday life from various causes. It is just to give some idea of the scale of risk, because I find that in public debate, for instance, people are not aware of the statistics, *and one has to keep them simple*. One has to provide the *minimum of information to get the message over*, and this Table I have found quite useful in this respect. [Emphasis added.]

This is a revealing admission. It runs counter to the views of the Royal Society Study Group and those academics whose empirical studies suggest that such comparisons are seriously misleading especially where nuclear power generation and radioactive waste are concerned. This approach also runs counter to the conclusions of the Royal Society's Study Group (1983, p. 143) that more attention should be paid to some confirmatory behavioural observations (revealed preference) and that the 'public's viewpoint must be considered, not as a form of indulgence or vote catching ... but as an essential datum'.

The Evidence of Professor Kletz

Professor Kletz is a specialist in the quantitative assessment of safety in the chemical industry, though he admits he is not an expert in the problems of nuclear industry (Kletz, 1984, p. 1). As to 'acceptable risk', he believes only a 'correctly informed' public (p. 22) should decide, and adopted a similar position to Matthews as to how this should be done.

> Industry should therefore inform the public of the relative size of various risks—*in ways the public can understand*—before doing what the public want. 10^{-6} means nothing to most people and even the scientist finds it hard to get a mental picture.

He followed this suggestion with a series of measures designed to equate risks, such as comparing equal probabilities, assuming that people could live for ever and calculating actual longevity for different exposures, calculating a shadow price on the cost of saving a life based on a value of life of £1 million per person, and listing out total numbers of people killed from various causes (Kletz, 1984, pp. 22–24). The explicit aim was to relate risks to common yardsticks—even though some risks were not so quantifiable as others (i.e. there is no sound historical record of serious nuclear incidents and no agreed data on radioactive induced cancer deaths). Also, Kletz concentrated on death, not delayed death or injury and made no mention of psychological stress, in which Eiser and his colleagues lay such store. He also observed that disproportionate amounts of money spent on saving statistical lives in the nuclear and chemical industries could be more effectively spent (in terms of saving lives generally) by raising the standards of living of the poor and in investing more in preventative medicine: an academic point, but one which is enormously difficult to put into practice.

Under cross-examination by Mr Brooke, Professor Kletz (Day 230, pp. 9–10) took a strong line on the matter of 'classifying' risk comparability for the lay public. He postulated that to choose a factor with a risk of 10^{-6} per year chance of causing death to a member of the public was akin to smoking $1\frac{1}{2}$ fewer cigarettes, or not drinking half a bottle of wine or four pints of beer per year. 'Admittedly', he noted (Day 230, p. 10B) 'these are voluntary and the other is not, but nevertheless, I think the comparison is fair ... If that is the only price you have got to pay it is hardly worth getting worked up and raising protest groups and getting all excited about a factory which is going to impose an equivalent.'

On the question of accident design targets, Kletz considered average risk, not the risk to the most exposed individual. He also used two parameters—namely total annual deaths and risk of death per year: both of these differ from the CEGB parameters (risk of death per person per year) (Kletz, 1984, p. 19).

> If everyone in a town of a million people was exposed to a risk of 10^{-5} to 10^{-6} per year from industrial hazards, the total number killed per year would average

between 1 and 10. *This would probably be unacceptable to Society* so an average risk of 10^{-7} has been suggested. This is the risk of being killed by lightning, a risk *most people* would regard as so low as they would not worry about it. [Emphasis added.]

On the matter of predicted failure (accident), Kletz was at pains to point out that few chemical plants contain safety systems of sufficient complexity to guarantee even a 10^{-5} per year accident rate. The one he could cite included 50 engineered trip indicators, each triplicated (Day 230, p. 16C). He believed that 10^{-5} was the best one could achieve for the totality of an engineered system. He proceeded to cast doubt on the accuracy and reliability of the 10^{-7} per year advanced by the CEGB for the design failure for the critical reactor pressure vessel (Day 230, p. 36D).

> There must be some point at which an estimated failure rate becomes so low that it is almost unbelievable for a man-made artefact and that unknown phenomena begin to creep in. I would not like to say without further consideration exactly at what level that occurs, but my feeling would be that when you get below 10^{-6} per year, the figure is so low that one has to be very cautious before accepting it.

Nevertheless, Kletz was quoted by FoE (Day 181, p. 68G) as believing that while 10^{-7} per year was acceptable risk for the average exposure, a figure of between 10^{-5} and 10^{-6} per day should 'perhaps' be the limitation on the maximum risk of death for the exposed population. Under cross-examination by Mr Brooke, Colin Green (Day 181, p. 68G) accepted that Kletz's observations provided a degree of support for the Board's target of 10^{-6}. In his cross-examination (Day 230, p. 46B) Kletz confirmed this view.

These two particular passages may well have convinced the Inspector that the 10^{-6} target level set by the Board was at least *responsible*, though not necessarily *achievable* (subject to as yet many unresolved safety clearances). However, these exchanges apply to *average risk*, not to risk to the most exposed individuals.

The Evidence of Dr Gittus

Dr John Gittus is Director of the Safety and Reliability Directorate of the Atomic Energy Authority, but was appearing for the Board as its witness on the techniques of PRA and, especially, of the probabilities of a loss of coolant accident (degraded core analysis). Irrespective of his position, he was asked by both the Assessor on nuclear safety (Professor Hall) and by the Inspector for his observations as to what criteria should be used to test for the competence of the PRA approach, to achieve reliability to the point of being scientifically reputable and, by inference, publicly acceptable (Day 214, pp. 56–62). Professor Hall was concerned about reliability of estimates when there are no experimental or statistical data. The Inspector asked what tests he should apply to all the safety evidence before the Inquiry so that he could advise the Energy Secretary that 'the public may rest content' (Day 214, p. 62B).

Following a weekend adjournment, and speaking from a prepared statement, Dr Gittus proposed that the Inspector should apply the three tests of *consistency*, *completeness* and *importance*. According to Dr Gittus, the level of consistency between data and analytical predictions on the one hand, and with experience on the other, ought to be established by the Inquiry. Second, the three elements of consistency, namely data, analytical capability and experience, should be examined according to their completeness. Third, a test of importance ought to be applied. Importance implied that not only the frequency and magnitude of the consequences of failure of a particular safety system should form a test of relevance to the Inquiry, but further, that lay and professional valuations as to importance ought also to form a part of the test. According to Dr Gittus (Day 215, p. 21D):

> ... the test of importance embodies, in my mind, the belief that public issues are more important at the Inquiry and to the public than professional ones.

But for the many detailed safety issues, Dr Gittus would further prefer a degree of professional autonomy (Day 214, p. 13B):

> The profession should meet its safety targets but to the extent that these may involve very detailed considerations such as clad ballooning and instrument air, they will be inscrutable to the public and are therefore (best left) to the professionals.

In other words, Dr Gittus' tests have the effect of confirming the *status quo*. Professional issues are those which are inscrutable to the public; public issues, i.e. *not* the detailed specialized issues, are those that should be dealt with by public inquiry. This response underestimates the need for an open examination such as the Sizewell B Inquiry to examine professional competence. As a result, where appropriate, 'professional' issues may become public ones.

Furthermore, it is revealing to examine the epistemological position which underlies such statements as those made by Dr Gittus, who was at pains to stress the following points to the Inspector.

(i) '... the scientific method is the method which has been developed over the centuries to supply (analytic capability). That is why so much of the Inquiry is concerned with science'. (Day 215, p. 19G);

(ii) degraded core analysis is a technique which exhibits analytic capability (Day 215, p. 19E);

(iii) '... degraded core analysis ... tells the public how the risks posed by Sizewell B compare with the risk they run in their everyday lives, and by implication accept'. (Day 215, p. 12G).

The import of this kind of reasoning is that the scientific method is capable not only of *determining* the level of risk, but also of *supplying the basis for the determination of acceptable risk* at a public inquiry. However, it can be argued that such positivistic determinations of publicly acceptable levels of risk are challengeable for their presumptions about the soundness of extrapolating from present circumstances to future impositions of levels of risk which are consequently deemed to be publicly acceptable.

Commentary

The tentative conclusion from our analysis so far is that the Sizewell B Inquiry seems rather too dependent upon scientific and professional judgement as to what should constitute acceptable levels of danger for the general public from a new nuclear reactor. This dependence appears to be a feature of the Inquiry itself. Its formal procedures based upon detailed testing of evidence and analytical cross-examination may have inhibited it from considering the wider social, psychological and political aspects of what constitutes acceptable risk.

We note that no psychologist or political sociologist was called as an independent witness on a par with Sir Edward Pochin and Professor Kletz, to comment on the profundity of the concept of 'acceptable risk', on the diversity of ways in which it should be discussed in a public inquiry, and on the manner in which the concept should be applied to engineered safety features, the licensing of the final design, the determination of a probabilistic modelling and presentation to the public.

Yet there is much much more to the Sizewell B Inquiry than the evidence we have covered. The cross-examination of Mr Matthews by FoE and the Inquiry Counsel suggest that the Inquiry may not be entirely satisfied with a single value for detriment (risk of death per person per year), and that comparative analysis of risks from different electricity generating sources, including other nuclear reactors are also being undertaken, *possibly with the use of other decision criteria*. The frequent reference to the Royal Society Study Group Report (1983), especially its well argued chapter 6, allows room for speculation here. In addition the Inquiry has exhaustively examined the procedures for setting and regulating safety in nuclear plants to be satisfied as to the competence of the institutions responsible for determining and protecting worker and public safety, and how they work, or propose to work should one or more PWRs be built.

The Sizewell B Inquiry is one of the most remarkable events of its kind. We would argue that it has genuinely attempted to understand the wider aspects of acceptable risk, and therefore has been thorough. The significant view expressed by the Council for Science and Society was that 'acceptable risk' is much more than a safety consideration: it has vital implications for emergency preparedness, local planning matters and negotiations with the affected community regarding such subjects as liability and compensation. Interestingly, all of these topics have been publicly canvassed through debate and analysis to considerable extent, and employing the very latest thinking at the Sizewell B Inquiry. We await the Inspector's Report and the subsequent reactions with interest.

Acknowledgements

This research is funded by a grant from the Economic and Social Research Council. The authors would like to thank Colin Green, Trevor Kletz and Roy Matthews for their comments on an earlier draft of this paper. The views expressed here are of course the sole responsibility of the authors.

References

Advisory Committee on Major Hazards (1976). *First Report*. London: Health and Safety Executive.

Ashby, E. (1978). *On Reconciling Man with his Environment*. Oxford: Oxford University Press.

CEGB (1982). *Statement of Case*. London: Central Electricity Generating Board.

Council for Science and Society (1977). *The Acceptability of Risks*. London: Barry Rose.

Department of Energy (1982). *Nuclear Power*. Cmnd 8317. London: HMSO.

De Piro, A. (1983). *British Gas Methane Terminal, Canvey Island*. London: Department of the Environment.

Fischhoff, B. (1983). Acceptable risk: the case of nuclear power. *Journal of Policy Analysis and Management*, **2**, 559–575.

Fischhoff, B., Lichtenstein, S., Slovic, P., Derby, S. and Keeney, R. (1981). *Acceptable Risk*. Cambridge: Cambridge University Press.

Green, C. (1983). Proof of Evidence: The Justification of the CEGB's Fundamental Reliability Criteria. FoE P/4 Sizewell B Inquiry.

International Commission on Radiological Protection (1977). Recommendations of the ICRP. *ICRP Publication 26*. Oxford: Pergamon Press.

Kemp, R. (1983). *Power in planning decision-making: a critical theoretic analysis of the Windscale inquiry*. Unpublished PhD Thesis, University of Wales, Cardiff.

Kemp, R., O'Riordan, T. and Purdue, H. M. (1984). Investigation as legitimacy: the maturing of the big public inquiry. *Geoforum*, **15**, 477–488.

Kjellstrom, B. (1983). Safety criteria and uncertainties in reactor fault transient analysis. FoE P/1 Sizewell B Inquiry.

Kletz, T. (1984). Proof of Evidence. The use of quantitative methods in risk assessment. TAK.P/1 Sizewell B Inquiry.

Macrory, R. (1982). *Public Inquiry and Enquete Publique: Forms of Public Participation in England and France*. London: Institute for European Environmental Policy.

Matthews, R. (1982). Proof of Evidence: The CEGB approach to nuclear safety. CEGB/P/2 Sizewell B Inquiry.

NRPB (1981). Cost benefit analysis in optimising the radiological protection of the public: a provisional framework. Chilton, Oxon: National Radiological Protection Board.

Nelkin, D. and Pollack, M. (1982). *The Atom Besieged: Anti-Nuclear Movements in France & Germany*. Cambridge, Massachusetts: M.I.T. Press.

NII (Nuclear Installations Inspectorate) (1979). *Safety Assessment Principles for Nuclear Reactors*. London: Health and Safety Executive.

NII (Nuclear Installations Inspectorate) (1983*a*). Proof of Evidence: H.M. Nuclear Installations Inspectorate's view of the CEGB's safety case. NII/P/2 Sizewell B Inquiry.

NII (Nuclear Installations Inspectorate) (1983*b*). Addendum 3 to NII/P/2. Sizewell B Inquiry.

NII (Nuclear Installations Inspectorate) (1984). Addendum 8 to NII/P/1. Sizewell B Inquiry.

Purdue, H. M., Kemp, R. and O'Riordan, T. (1984). The context and conduct of the Sizewell B Inquiry. *Energy Policy*, **12**, 276–281.

Royal Commission on Environmental Pollution (1976). *Nuclear Power and the Environment*. Cmnd. 6618. London: HMSO.

Royal Society Study Group (1983). *Risk Assessment*. London: Royal Society.

Starr, C. (1983). Coping with nuclear power risks: a national strategy. In Covello, V. T., Flamm, W. G., Rodricks, J. V. and Tardiff, R. G. (eds), *The Analysis of Actual versus Perceived Risks*. New York: Plenum, pp. 251–258.

van der Pligt, J. (1985). Public attitudes to nuclear energy: salience and anxiety. *Journal of Environmental Psychology*, **5**, 87–97.

Webb, G. A. M. (1984). The Requirement to keep Radiation Exposures as Low as Reasonably Achievable (ALARA). NRPB/P/3 Sizewell B Inquiry.

Wynne, B. (1982). *Rationality and Ritual: The Windscale Inquiry and nuclear Decisions in Britain*. Chalfont St Giles, Bucks: British Society for the History of Science.

INDEX